History of Life Sciences

and Medicine 2

SCIENCES

History of Sciences, Field Director – Jean-Claude Dupont

Life Sciences, Subject Head – Jean-Claude Dupont

History of Life Sciences and Medicine 2

Modern Era (16th to 18th Centuries)

Coordinated by

Céline Cherici

iSTE

WILEY

First published 2026 in Great Britain and the United States by ISTE Ltd and John Wiley & Sons, Inc.

ISTE Ltd
27-37 St George's Road
London SW19 4EU
UK

www.iste.co.uk

John Wiley & Sons, Inc.
111 River Street
Hoboken, NJ 07030
USA

www.wiley.com

Library of Congress Control Number: 2026931545

British Library Cataloguing-in-Publication Data
A CIP record for this book is available from the British Library
ISBN 978-1-78945-209-9

ERC code:
SH6 The Study of the Human Past
SH6_14 History of science and technologies, environmental history

The manufacturer's authorized representative according to the EU General Product Safety Regulation is Wiley-VCH GmbH, Boschstr. 12, 69469 Weinheim, Germany, e-mail: Product_Safety@wiley.com.

Contents

Presentation of Field "History of Science"

Jean-Claude Dupont
CHSSC, Université de Picardie Jules Verne, Amiens, France

The "History of Science" field of the *SCIENCES* collection is closely related to the sections devoted to scientific and technical disciplines, on the one hand, and to epistemology, on the other hand. In addition to providing a summary of current methods and issues in the history of science, all of the volumes offer an overview on the evolution of knowledge in mathematics, the physical and natural sciences, medicine, the humanities and technology, from the beginning to the present day, as well as in different cultural areas. This is aimed at non-specialists and scientists alike, providing them with a useful perspective on the origins and foundations of scientific knowledge, while keeping up with contemporary research in the history of science.

– Field Director: Jean-Claude Dupont.

– Subject Heads of the English edition:

- "Issues, General Themes and Methods of History of Sciences": Stéphane Tirard.

- "Logic and Mathematics": Sabine Rommevaux-Tani.

- "Astronomy, Physics, Chemistry and Earth Sciences": Christian Bracco.

- "Life Sciences": Jean-Claude Dupont.

- "Humanities and Social Sciences": Alexandre Escudier.

Introduction

Céline Cherici
CHSSC, Université de Picardie Jules Verne, Amiens, France

In his 1977 book, *Ideologie et rationalité dans l'histoire des sciences de la vie* (*Ideology and Rationality in the History of the Life Sciences*), Georges Canguilhem asserted that the history of science is not just the memory of science but also the laboratory of epistemology. If this is the case, then we believe that the history of medicine goes hand in hand with an epistemological and critical perspective on developments in the biological and medical sciences. Thus, navigating through the modern era requires complex reflection, the deepening of a wide range of knowledge and the correlation of scholarly discourses, which are sometimes traditionally regarded as antagonistic. A conceptual study of the history of the life sciences from the 16th to the 18th centuries helps us redefine the links that structure the technical, institutional and scientific relationships between medicine and biology in a period when the understanding of bodies fluctuated between several models, ranging from mechanism to the consideration of matter in motion, driven by hierarchical functions.

A study of the disruptions, obstacles and recurrences that have marked these two centuries leads us to reflect on unique ancient concepts such as "life, anatomy, experience" within the context of particular works and contexts. This perspective of a history recorded through selected examples allows us to question whether the science of the past is the same as the science of the past that we know today or whether it possesses unique traits that make it a multidisciplinary history that is perpetually renewed. The history of the human body, and of the place of animals and plants in the history of our understanding of the world, is polymorphous, numerous and pluri-methodological, with as many techniques, imaginations and experiments as there are scientists. From the opening of bodies to the understanding of diseases and from the observation of clinical symptoms to the exploration of internal structures, there is a whole world in which functional and organic normality and pathology have been defined and objectified, which needs to be understood. Dynamic thinking and models intertwine with one another: such is the example of clinical medicine and anatomopathological medicine, which are based on comparative anatomy as much as methods for classifying animal and plant species. This volume encouraged us to explore the 16th, 17th and 18th centuries in this way. The structure of the various chapters emphasizes the relationship between this chronological section, the texts studied and the analysis of trends, representations and techniques over a relatively long period, during which time medicine took on numerous models in the European cultural area.

At the 12th International Congress on the History of Science, held in Paris in 1968, Suzanne Bachelard underlined the fact that while the historian's activity, being based on the analysis of the past, imposes limits on him, it also allows him to reconstruct his object in a chosen time and space. Several levels of analysis intersect to identify the foundations and conditions of emergence and development of scientific concepts and objects: the tools and methods used by scholars, the theoretical concepts, the ways in which knowledge is disseminated (learned societies, journals and memoirs) and the types of discourse used. These elements provide a framework for epistemological analysis, enabling us to trace the development of biological and medical knowledge.

Although medicine can be divided into four fields of observation – physiology, pathology, prophylaxis and therapeutics – conceptual unity is not always achieved. From the work of Vesalius to that of Morgagni, we will analyze medical theories that were considered antagonistic but that overlapped, complemented each other and contributed to the renewal of the medical art. The scholarly, empiricist and mechanist schools, whether in Galenic and Aristotelian tradition or, on the contrary, marked by a desire to innovate, focused on developments in scale, techniques (microscopies) and perspectives. Throughout the 17th century, the success of the mechanistic approach to the phenomena of life intersected with vitalism, and, to some extent, overlapped with it. While Aristotle's theories seemed to have been toppled by the Cartesian mechanism, the various trends studying the living world mutually enriched and integrated with one another. Until Leonardo da Vinci (1452–1519), there were two main "camps", which were less opposed than they might have appeared: on the one hand, the Galenists, with their various theoretical and practical variations, and, on the other hand, the Aristotelian followers. Their respect for the school of Hippocrates was their common ground. Until the 17th century, almost all biologists and physicians were marked by a conclusive view of nature and the bodies that make it up. This was followed by the development of innovative and exploratory medicine, which, while asking "how", continued to ask "why" and "to what end". In the history of biology, conceptualizations of phenomena of structural and functional normality in living beings, which a posteriori can be described as pre-scientific, have often been extended, in different forms, as necessary elements for the development of seemingly new theories or disciplines. The theory of organic incitability, based on the concept of vital force, marked by the movement of life between internal forces and external agents, questions both "how" and "why". Brown, for whom life is that which is constantly destroyed, encounters the ideas of electricity and converges with Bichat's perspective, who, a contrario, defines life as that which resists death.

Understanding and describing these nuances, to account for the complexity of developments in the life sciences, requires a conceptual and thematic approach, insofar as this knowledge is reconstructed in terms of the particular issues and examples. Highlighting the interweaving of knowledge, but also the faults that punctuate its development, prevents us from mistaking the persistence of terms for concepts or mistaking similar descriptions for analogous theories. An epistemology of disruptions, divided into chapters, enables us to focus on more or less short periods of time while approaching a two-century period marked by a mix of currents of thought and many developments.

From the 16th to the 18th centuries, all diseases were typically reduced to alterations in vital movements and/or organs. During the Middle Ages, Galenism was enriched by a neo-Hippocratic movement, partly resulting from the rejection of Galenism in university teaching, but also from feedback from bedridden subjects, and the growing importance of the scholar's gaze at anatomical structures and symptoms, a gaze that creates links and progressively constructs its object.

The exploration of the living world therefore took many different paths, complementing anatomical knowledge at the time, exploring what could be observed with the naked eye (micrography) and attempting to classify the new living forms that were discovered, owing to micrography and the development of taxonomy. The recurrence of scientific terms in a discipline should not obscure the fact that the objects described differ from one period to another: the cell described by Hooke in the 17th century is not comparable to its later developments. However, it is also a question of understanding the functions that correspond to each of the anatomical forms described through physiology and of learning about their developmental foundations through embryology. The 16th and 17th centuries in Europe thus opened new perspectives in the field of living organisms, not only from a methodological and technical point of view but also from a conceptual one.

Explorations of the organism multiplied at various levels, from internal structures to organs, tissues, nerves and cells. The scholar's gaze became multifaceted, considering the intersections between electricity, chemistry, physics and medicine. New models were developed to represent the different forms of living organisms, both healthy and sick, from anatomy to clinical pathology. The categorization of animals, which dates back to Antiquity, was developed through taxonomy, with the aim of understanding how animal populations are distributed and how they can be grouped into hierarchical groups according to their anatomical and phylogenetic affinities. The notion of "partial evolution" first appeared in Kircher's writings as early as 1675 and was further developed in comparative anatomy, whose origins go back to Antiquity. However, the idea of the existence of evolution among species appeared through Maupertuis and Buffon's research in the mid-18th century. Maupertuis suggested that the first forms of life arose by spontaneous generation from random combinations of molecules or germs. However, he presents each species a priori as clearly distinct from its neighbors and does not consider the natural selection of modified traits to justify their development. The development of these ideas makes it possible, among other things, to conceive of similarities between animal organisms, such as the donkey, the zebra, the horse, the lion, the lynx and the cat; to imagine subdivisions within the same category and to consider processes of transition from one form to another. From the latter to transformism, this step would only be taken at the end of a long societal, cultural and anthropological journey, as well as a scientific one. Similarly, in the field of botany, Linnaeus remained committed to the fixity of species, even as he was overcome by doubt in the mid-18th century. In fact, by using the vocabulary of genus and variety and by asserting the possibility of links that allow a genus to change into secondary species, he admits that the various species can be derived from a common strain.

As the result of a complex development, the various models used to classify and establish the anatomical, physiological and behavioral profiles of the main species originated from the Renaissance period and were essentially expanded on in the 18th century. The observation and classification of plants during the modern era corresponded to several research programs. Discussing developments in botany during the Renaissance requires us to consider not only classical texts from Antiquity, but also new explorations and descriptions of the plant world, as well as new methodological and technical tools (printing and distribution of books) at the foundation of the discipline's renewal.

This journey to the foundation of the history of science and medicine in the modern era will take us from *dissection to the site of diseases*: in other words, the 16th–18th centuries. Céline Cherici will highlight the key stages in this journey, which explores the body to heal a human being. We will continue this research by exploring *clinical practice or the art of diagnosis* through the analyses of Jean François Thurloy. He will present some of the founding works of the *clinical approach*. Jean-Claude Dupont will look at how *experimentation on living beings* in the modern era was formed by new technologies, which in turn shaped new medical and physiological perspectives. Paolo Quintili will present models of *representation of living beings*, taking us on a journey of discovery through *images of the body* and the links between the microcosm and macrocosm. Jean-Luc Guichet will place knowledge of *animals* in the broader context of the *observation and classification* of living beings. Caught between *natural history, taxonomy, zoology and ethology*, knowledge of animals is built around as many scales as discoveries. Olivier Perru introduces important aspects of the *history of the plant*. From the 16th to the 18th centuries, scholars struggled with different *models of the living being*, and their conceptions of *plants* were sometimes at odds with one another. Paolo Quintili will detail various principal *systems of the living being*, from *Descartes to Barthez*, showing the extent to which metaphors and elements of *mechanicism, iatrochemistry* and *vitalism* constitute them. Finally, Stéphane Tirard will conclude the volume with an in-depth study of the emergence of *transformism in the 18th and 19th centuries*.

References

Bachelard, S. (1968). Conditions et progrès de l'objectivité en histoire des sciences. In *Volume 1/A du XIIe Congrès international d'histoire des sciences. Colloques : textes et rapports*, Paris.

Canguilhem, G. (2009). *Idéologie et rationalité dans l'histoire des sciences de la vie*. Poche, Paris.

Grmek, M.D. (1995–1999). *Histoire de la pensée médicale en Occident*, 3 volumes. Le Seuil, Paris.

Morange, M. (2008). *À quoi sert l'histoire des sciences ?* Quæ, Versailles.

1

From Dissection to the Origin of Diseases: 16th–18th Centuries

Céline CHERICI
CHSSC, Université de Picardie Jules Verne, Amiens, France

1.1. Introduction

Between the 16th and 18th centuries, a medical and philosophical tension existed between the cartographic aspect of human anatomy, facilitated by the advent of printing, and the development of clinical thinking. The 17th century was marked by major biomedical discoveries, such as the circulation of blood and the development of dissection and body preservation techniques, but also by controversies over the terms used in human anatomy. The 18th century, however, was marked by technical and medical breakthroughs, which appeared amid controversies between innovators, supporters of flawless medicine and researchers of organ groups and tissues.

By remaining focused primarily on the morphological and "geometric" aspects of disease, pathological anatomy tends to disassociate organic traits observed in clinical practice, which is itself rooted in the time frame of pathological processes. This chapter's objective is to shed light on some of the trends that shape medical practice and, through the perspective of the conceptual history of medicine, to analyze how scholars have explored the body, its ailments and its mechanisms since the Renaissance. Anatomic pathology is essentially polymorphic in its iconographic and textual representations and, since its inception, has marked an expanding knowledge of the human body. Between the 16th and 18th centuries, a medical and intellectual trend was observed, shifting from text to observation, from books to the body, from narration to flesh, and extending to "[…] the art of reproducing the open body in images, including its organs, its tissues and its membranes" (Adisson 2018, p. 34), thus involving direct dissection.

The Renaissance saw an increase in dissection activity, making it possible to systematize the direct examination of the state of corpses in places where bodies were found under varying conditions. Owing to his illustrations, Vesalius laid the empirical foundations for the famous maxim: "Open up a few corpses: you will dissipate at once the darkness that observation alone could not dissipate" (Bichat 1801, Foreword).

Thus, from observing the body in motion to observing clinical symptoms and exploring organic structures, it is necessary to understand the boundaries between normality and pathology. Dynamic medical ideas meet and interweave with one another: this is the case with clinical medicine and the anatomopathological method. Medicine is divided into four fields of observation, forming a body of knowledge that continues to evolve: physiology, pathology, prophylaxis and therapeutics. Knowledge of physiology intersects with pathology, which sheds light on it as well as on prophylaxis and therapeutic approaches.

First, through autopsies and dissections, we consider the extent to which new explorations of the human body can link cuts and the search for signs of disease to one another. We analyze several elements of the history of the development of a Cartesian mechanistic model, questioning how it leads to a new dismantling of the body into sets of structures. Finally, we will explore the impact of anatomical research on clinical thinking, particularly through some aspects of the work of Giovanni Battista Morgagni (1682–1771). In this work, we will see how trends in medical exploration and theory, which are considered antagonistic, intersect, complement one another, and renew the art of medicine.

1.2. New explorations of the human body

In the 16th century, the discovery of new landscapes was seen as a source of observation and experience. This perspective was infused with anatomical investigations. While many future practitioners were engaged in studying ancient texts from the Hippocratic corpus, Aristotle and Galen, students were enthralled by the dissection demonstrations performed by Andreas Vesalius (1514–1564)

(Klestinec 2004, p. 376). Vesalian anatomy can be seen as the culmination of an anatomical renaissance that began in the 14th century with the development of Italian and French universities. Its significance should neither be downplayed nor consolidated. Far from being homogeneous, the research of Renaissance scholars contributed to the formation of new medical knowledge. We must avoid simplifying the difference between learning from books and sudden epistemological disruption. While continuing to study and teach the practical and philosophical aspects of anatomy, which is based on an ancient textual tradition, anatomists have focused their research on the faculties of the human body and on understanding the limits or permeability between disease and health. This approach encourages the emergence of cross-disciplinary practices involving reading, dissection and direct representation of the structures observed. Dissection methods, which vary according to the targeted organs, involve methodological and theoretical challenges. Thus, the renaissance of anatomical science was facilitated through the linking of texts from classical authors with the emergence of new perspectives, both anatomical-pathological and clinical. A succession of scientific discoveries made this merging of ancient texts with exploratory practices subtle and complex. With respect to continuity and discontinuity, anatomy gradually became the foundation of clinical and physiological approaches. Thus, to shift from searching for the causes of death to the scholarly exploration of anatomical structures, it was necessary not only to rethink dissection techniques and combine them with new preservation techniques but also to combine the study of healthy bodies with that of sick individuals, as well as those who had died from disease, with a focus on organic damage in correlation with the physical aspects of health. This movement toward understanding the body took root in the context of medical autopsies and evolved into scholarly dissection during the 16th century.

1.2.1. *A renewal of anatomy led by Andreas Vesalius (1514–1564)*

As early as the 15th century, Jacopo Berengario da Carpi (1447–1530), a professor in Bologna, claimed to only trust the evidence experienced through his senses, drawing on his empirical practice of medicine to develop his teachings on anatomy and surgery. The University of Bologna played a fundamental role in the development of anatomy. In the mid-16th century, students expressed their desire to learn structural anatomy (Klestinec 2004). As a result, Vesalius, who had been in Padua since 1537, was invited to develop demonstrations for the lectures of Matheaus Curtius (1475–1542). The students flocked to his demonstrations. In the preface to his *De humani corporis fabrica* (Vesalius 1543), Vesalius unequivocally declared the advent of an innovative anatomy on the basis of the skill of the hand and the eye and on new methods in medicine. He challenged a form of medicine that was far removed from anatomical exploration[1], describing the paradox of an art that seeks both knowledge and healing:

> [...] it is clear that, of all that which has been conceived by human genius, the art of medicine is the most useful, by far the most necessary, but also the most difficult and demanding; nothing more disastrous could have happened [...] than the moment when medicine began to be fragmented and its principal instrument of treatment, manual surgery, was neglected to the point of being seemingly abandoned to common people and to individuals who had received no instruction in the disciplines of the art of medicine. (Vesalius 1543, Preface)

While Charles Estienne (Estienne 1545, p. 343) regarded the eye as the only organ that could guarantee the accuracy of knowledge, William Harvey (1578–1657) based his treatise on the circulation of blood (Harvey 1628) on autopsy and experimental findings. Jean Riolan Younger (1577–1657) said that he worked with "ocular hands" (Riolan (the Younger) 1653, p. 26), and many mentioned the influence of the senses on the theoretical practice of medicine, which was firmly rooted in the exploration of the body.

Observations made during autopsies were combined with the teaching of human anatomy and played an increasingly significant role in the development of medical education. In an attempt to determine the cause of death, autopsies became educational and a source of knowledge. As a privilege of Italian university towns, they enabled the identification of specific pathological conditions, thus providing a framework for understanding diseases. The art of medicine was displayed in anatomical theaters and was drawn and depicted in illustrations, similar to something between cartography and the formation of knowledge.

In Padua, where the study of anatomy was linked to the teaching of surgery, Vesalius systematized the representation of the structures of the human body. He thus sought to teach the organic reality of bodies in all their states. In the context of the controversy between anatomical terms and the realities they refer to, the

1 "Indeed, while anatomy was once held in high esteem by physicians, who devoted all their efforts to mastering it, it began to decline miserably when physicians abandoned its study and referred surgical operations to others [...]. And from the moment doctors left the entire management of dissection to barbers, they not only lost their true knowledge of the viscera, but also their skill in dissection" (Vesalius 1543, Preface).

anatomist's task was to explore and depict the unknown landscapes referred to in ancient texts:

> Furthermore, so that my books may be of particular use to those who do not have the opportunity to examine [a dissected body], they describe the number, location, shape, size, and substance of each part of the human body in sufficient detail, as well as its connection to other parts, its usefulness, its function, and many other characteristics of this kind that we are accustomed to examining in the nature of the parts during a dissection; they also provide information on the procedures for dissecting dead and living bodies and contain images of all parts of the body inserted into the text in such a way as to place, so to speak, a dissected body before the eyes of those who study the works of nature. (Vesalius 1543, Preface)

Topology combines methodology and pedagogy with the aim of acquiring empirical knowledge. For example, Chapter VII of Book VII of *Fabrica* shifts toward an anatomical and physiological description of the brain, comparing it to a *turtle shell*[2] for those new to anatomy. The aim is to use what the anatomist sees as the descriptive touchstone of physiology. Morphology thus explains structures in the body.

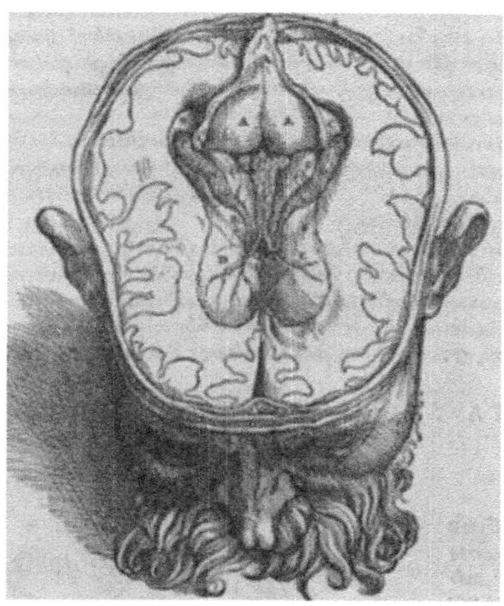

Figure 1.1. *This diagram illustrates sections of the brain, showing the first cartographic elements (Vesalius 1543, Fig. 6, Book VII, p. 611)*

2 By supporting the weight of the bodies placed above it and preventing them from falling, this *turtle shell* protects and prevents the large cavity of the third ventricle from collapsing.

Vesalius systematized dissection to develop therapeutic approaches and facilitate the learning of surgery. Surgeons must first train on cadavers to more easily approach living beings when they want to study the action and function of body parts (Vesalius 1543, p. 658). While animal vivisection may help in understanding the functions of the human body, it cannot produce identical results, which means that both human dissections and illustrations are necessary. From learning about anatomy to surgical care techniques, students move from animals to humans and grasp the complexity of organ functions, such as the liver, kidneys and bladder. From the study of living tissue to *postmortem* structures, from autopsy to knowledge, sources of innovation and representation, direct anatomy has established itself as a specific discipline. Vesalius made it possible to forge links between speculation and *de visu* investigation, archived through heuristic diagrams. *La fabbrica del corpo umano* (Vesalius 1543) contains approximately 300 of these, including extraordinary diagrams of the brain that allow him to differentiate between white and gray matter, as well as the corpus callosum. The theoretical idea was to look at his illustrations as if looking at a dissected body. Padua remained a center of anatomy for a long period of time. Realdo Colombo (1516–1559), Gabriele Falloppia (1523–1562) and Giroldamo Fabrici d'Acquapendente (1533–1619) made notable discoveries in physiology. Colombo performed numerous dissections, correcting existing knowledge, whereas Falloppia published *Observationes anatomicae* in 1561. D'Acquapendente (1603) founded the anatomical theater of Padua in 1594, designed a comparative anatomy atlas in color, used compression straps for bloodletting and demonstrated the valve system of veins. The medical centers of Montpellier, Basel and Paris also developed into major centers of anatomy. Paris was considered the fortress of Galenic knowledge, and the work and research of Jacobus Sylvius (1478–1555), Vesalius' critic, dominated the scene. However, his medical school also saw the publication of Estienne's research, which criticized Galenic anatomy for being indirect, as well as Ambroise Paré's (1510–1590) universal anatomy in 1561, in which the surgeon helped spread Vesalius' ideas. Jean Fernel (1506–1558) defended the idea that anatomy was as essential to medicine as geography was to history. In 1542, he published *De naturali parte medicinae* and coined the term "physiology". He described it as inseparable from anatomy but distinct from pathology. Although his work was seemingly moving toward a solidist physiology, he remained a proponent of humorism while referring to Galenic explanations of localization. While the anatomy disseminated by Vesalius was part of a broader epistemological break in medical practice, we should emphasize the heuristic perspectives derived from autopsies, which were widely practiced in Italy in the context of criminal investigations.

1.2.2. *From autopsies to knowledge of pathologies*

Between a clinical perspective and a humoral pathology, anatomy pays careful attention to the philosophical aspect, particularly by focusing on the study of Aristotelian principles. The theoretical encounter between the works of Galen and Aristotle, combining descriptive anatomy with philosophical thought, formed an initial dynamic of normal and pathological states of the body. These texts frequently overlap in the research of many scholars and offer opportunities for empirical exploration of the body. A debate illustrating some of the paradoxes of medicine, caught between tradition and innovation, pits two cities in Sicily against each other, namely, Messina and Catania. This example, taken from the microhistory of medical science, highlights the gap between highly detailed pathological descriptions and the inability to treat diseases that are well described: this may be because of a lack of direct anatomical knowledge. In this dispute, which illustrates the high-profile debates in medicine between the 17th and 18th centuries, two schools of thought emerged: a conservative model known as "veterinary-biliary" and a "neoteric" trend[3], notably supported by Marcello Malpighi (1628–1694) and Giovanni Alfonso Borelli (1608–1679). The influx of new scientific ideas complemented traditional perspectives and questions about humoral physiology (Montoliu 2021, p. 7). While knowledge about diseases came about owing to developments in anatomy, this knowledge became deeply embedded in the tissues, internalized through this great open book on organic structures.

Scientists sought to highlight these origins from the beginning of the 16th century, in the context of autopsy-related practices. Although these autopsies did not initially have a scientific purpose but rather a causal perspective on death, these observations nevertheless played an increasingly important role in the development of medical learning.

In the Florentine autopsies regularly published in the works of Antonio Beniviani (1443–1502), on deceased individuals whose *intra vitam* problems he was familiar with and whose hidden causes he sought to uncover, as well as in Fernel's treatise *Universa Medicina* (Fernel 1567) or that of Volcher Coiter (1534–1576) (Coiter 1566), anatomical exploration is taken as the guiding principle of their research. Adherence to a certain Galenic tradition is perhaps consistent with the organization and rationalization of the nomenclature of anatomical descriptions. The scholars who preceded the "Vesalian revolution" referred to pathological observations and contributed to the development of anatomical investigations that focused on

3 The adjective *neoteric* is used to describe innovative and/or exploratory medicine, often detaching itself from medical approaches based on ancient texts (Roger 2014, p. 123).

different objectives: determining the causes of death, defining pathological forms and their effects, and understanding anatomical norms. This last point can be considered a consequence of the previous two. In the 16th century, an epistemology that promoted the integration of anatomical pathology as a specific field of knowledge emerged. This body of work, which was between textual traditions and physiological innovations, was part of the context of humoral pathology that was already linked to localization:

> In their eyes, anatomical knowledge was of immediate interest to medical practice, as many disorders and diseases had their specific place in the human body or could be attributed to a specific organ. (De Renzi 2019, p. 80)

Interest within the organic context of diseases is an element of humoral medicine. Italy is the center of a field of anatomy where the boundaries between the exploration of the healthy body and the search for morbid processes intertwine with one another. Anatomists attribute meaning to this intertwining of normal and pathological explorations without breaking from the context of humoral pathology. While anatomical dissections are exploratory, autopsy investigations serve to search for causes and pathological forms. The field of the former needs to be structured according to the acquisition of knowledge and beyond a single practical purpose. Morbid explorations must be separated from criminal explorations to better focus on knowledge of diseases and the causal or secondary traces they leave behind. The complementary disciplines of anatomy and pathology work hand in hand to construct a scholarly framework for understanding the human body. The success of Falloppe's Paduan dissections, known as *sector*, was such that the students demanded the exclusion of Vittore Trincavella (1496–1568), who was a *lector* at the time, so that the teaching of anatomy could be entrusted entirely to him (Soulier 2021, p. 6).

Scholars have also developed a clinical eye to understand the structures of diseases in vivo, and after death, they have conducted anatomical and pathological investigations. Therefore, in this context, can humoral pathology be considered an epistemological obstacle? Nothing is less certain, given that the concept of localization is intertwined with the model of humors, which are themselves localized. If we question how to find a place for *postmortem* lesion observations in the form of medicine that attributes the causes of disease to the movements of morbid humors rather than to structural organic processes, we should consider the fact that humors are substances that depend on organs. Humoral pathology seems to involve not only notions of imbalance and equilibrium but also the movements of organs within the body. In the 16th century, it was often the malfunction of part of an organ that was highlighted to explain a death, implying a localization-based view rather than a holistic perspective. Until the early 17th century, among Galenic physicians, ailments were attributed to corrupted substances, which showed signs of pathogenic qualities triggered by the quality of the air or food[4].

These substances are digested in the stomach, and if the body is unable to transform them into blood or eliminate useless elements, they stagnate, rot and cause morbid substances to accumulate in specific organs, thus causing poisoning. All this takes place within organic spaces that can be explored and mapped and therefore located. Galen introduced the concept of *locis affectis*[5]. At the clinical level, the symptoms and signs of disease already indicate a site dependent on the organ from which they originate. A long tradition of the concept of localized disease is also evident in the terms used: lethargy refers to the brain, syncope to the heart and pleurisy to the lungs.

In autopsy reports, numerous descriptions of local alterations are used as an explanatory framework for understanding the links between diseases and subsequent death. For example, Rembert Dodoens (1517–1585) reported a case of aneurysm rupture (Dodoens 1581, p. 123), in which he observed a small amount of blood flowing from the mouth. After investigation, the scholar found a large pocket of blood on the left side of the thorax. Although not easily visible from the outside, this pocket was revealed by a localized search for the possible cause of death. Ercole Sassonia (1551–1607), a professor of practical medicine in Padua, noted a visibly altered substance in the lungs but no external noise that would indicate this morbid process (Sassonia 1607), which nevertheless appeared to correspond to a case of dyspnea. In these examples, pathological accumulations of substances are noted without having previously caused any significant series of symptoms. Dissection alone made it possible to identify the possible sites for these morbid processes. Thus, numerous autopsies have made it possible to identify pathological humors, both liquid and solid, in the process of forming pathological masses. Observations of scirrhus, stones and tumors suggest changes that directly reflect different morbid dynamics in the body.

---

4 "If, in all the corpses he dissected, he always found damaged lungs in those who had pleurisy, he must consider that he observed this in Rome, where, due to the variability of the climate and excessive humidity, most diseases originate in the head, and it is therefore not surprising that the pleurisy and perineumonia that are widespread here are almost all of this variety" (Marziani 1626, pp. 78–79).

5 The treaty of the same name, which is entirely devoted to showing how to make accurate diagnoses, refers to the idea of locating the disease.

Collections of autopsy observations shape medicine while attempting to establish norms for bodily functions. By drawing inspiration from dissections, physicians could hope to improve their therapeutic results, insofar as *postmortem* signs offered hope of tracing symptoms back to their source. A science of correlations was developed between clinical and autopsy signs, in line with medical practice. The importance of sight and touch in the act of dissection itself grew alongside the development of clinical practice. Since identifying the internal causes of disease is considered the foundation of therapy, dissecting bodies became fundamental to improving patient health. The increase in dissections during the 16th century gave rise to new practical knowledge that accompanied the reading of classical texts. Through one exploration after the other, scholars classified, described and located the signs of disease to recognize and treat them.

1.2.3. *Exploring the signs of disease*

Classifications of clinical symptoms were developed prior to Morgagni's work, notably by Thomas Sydenham (1624–1689), who combined clinical considerations with pragmatic therapeutics. He was confronted with Jean-Baptiste van Helmont's (1579–1644) model of archaic physiology and Borelli's iatro-mechanical research program. He started from the principle that diseases become visible through grouped symptoms that form a picture of the disease. He thus sought regularities in symptoms and developed a dynamic clinical approach:

> Sydenham first sought to define the specificity of medicine in relation to natural philosophy. Medicine is the discipline that deals with 1) studying the natural history of diseases and 2) finding specific remedies for each disease. [...] Sydenham therefore argued that the art of medicine could only develop through the compilation of the history of diseases and continuous praxis, in other words, the practice of treatment. (Zampieri 2016, p. 133)

Although he failed to link the symptoms to anatomical changes, in his *Treatise on Gout* (Sydenham 1683), he accurately described the relationship between the patient's lifestyle and the attacks of the disease, from their paroxysm to their chronic nature. His descriptions, from a neo-Hippocratic movement, painted an understanding of diseases. Thus, in the 1882 translation of the *Treatise on Gout*, he correlated the poor diet of those affected by this condition and their rather sedentary lifestyle with their constitution and history of symptoms. The treatise concludes with "And thus unfolds the history of the disease" (Sydenham 1882, p. 19). This short sentence marks the end of the text, as it would in a short story, a story that recounts the clinical features of a disease, its place in individual biographies and its effects on the body's functions. This rhetorical process highlights the importance of narrative in understanding the characteristic signs of diseases.

In 1679, Théophile Bonet (1620–1689) published *Sepulchretum*[6], a work of major significance for the development of anatomical medicine. This treatise is a large anthology of *postmortem* observations of the organic signs of disease. Its subtitle immediately announces its programmatic perspective: *Practical anatomy based on corpses that died of disease. This paper presents the case history and observations of practically all diseases of the human body and reveals their hidden causes. It is appropriate to call this work the foundation of true pathology and correct nosology, or even the guide to ancient and new medicine. With the necessary indexes. A work long desired by scholars of medicine and anatomy was collected and compiled with great effort.* It thus presents itself as a methodological tool for conceptualizing direct observations that have been made on the human body and correlating them with known diseases. This is certainly what the cover illustration suggests, showing Bonet holding a pen and a notebook, which directly suggests the post-Vesalian approach to reading the body, narrated and illustrated by anatomical representations.

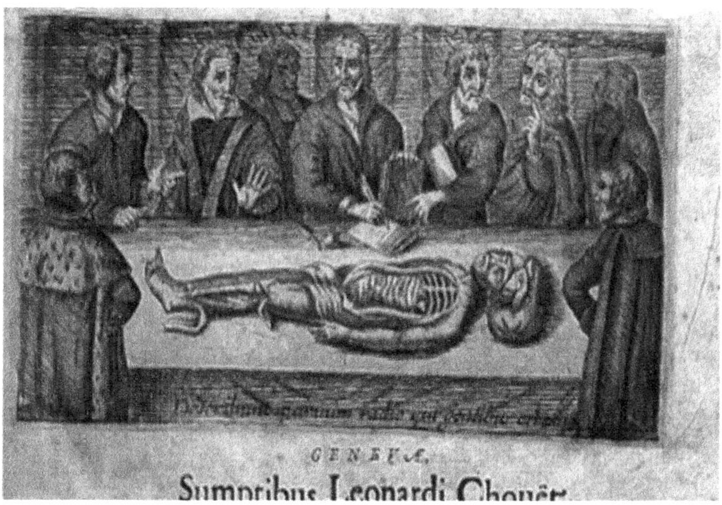

Figure 1.2. *Miniature art illustrating the cover of the* Sepulchretum *treatise (Bonet 1679)*

6 The treatise is dedicated to Thomas Bartholin (1616–1680), who played an important role in research on the lymphatic system (Sydenham 1653).

By collecting accounts of dissective and clinical observations, Bonet organized the lesions that were theoretically referred to into a single model, which was intricately understood in terms of its pathogenic mechanics. His compilatory approach aimed to reorganize fragmentary information and disseminate the concept of practical anatomy, which had already been formalized in 1600 by André du Laurens (1558–1609). The latter used it to identify what is operational in anatomy, as opposed to what is intellectual. It was also developed by Jean Riolan (the Younger) in *Anthropographia* in 1618 and in 1653 in his treatise *Manuel anatomique et pathologique*, in which the term *practical* appears 70 times. Hippocratic medicine is intertwined with an understanding of physiology on the basis of anatomical structures. While accepting Galenism, Jean Fernel (1506–1558) expressed a desire to reform its study, which he understood to include the study of the causes of diseases and their symptoms. In his *Thérapeutique universelle* (Fernel 1655), in Chapter IX, devoted to bloodletting, he stipulates that the strength of the disease should be judged by the appearance of the affected part, "[…] rotten, or stained with some pernicious quality […]" (Fernel 1655, p. 84). The cause of the illness should be discovered by "[…] examining the vessels, viscera, and the rest of the body […]" (ibid.).

Since the Renaissance, physicians[7] have produced knowledge about diseases on the basis of dissective anatomy. The genealogy of practical anatomy lies between pathology studies and anatomical investigations. In the 1670s, the relationship between truth and medical knowledge was the subject of intense debate, giving rise to relentless criticism of the conjectural nature of medicine (Rinaldi 2018, p. 50). Bonet thus marked the future developments of practical anatomy: from a tool for investigating individual cases and an occasional method of medical inquiry, he called for it to become a complex and coherent system on the basis of technical procedures and profound reflections on medical practice. By compiling his work into a corpus of shared knowledge, he put practical anatomy in the position to reorganize the various approaches to it.

Developments in the practice of dissection during the 17th century followed a path that was somewhere between autopsy and exploration. In 1666, Malpighi created a circle of dissectors (Ladislao 1966, p. 193), whose members considered pathological dissection to be part of the identity of medical art. They also performed

7 In addition to those already mentioned, we can mention the works of Guillaume de Baillou (1538–1616), Daniel Sennert (1572–1637), Felix Platter (1536–1614) and Willem Piso (1611–1678).

autopsies in Bologna hospitals as part of their institutional duties. Malpighi used *postmortem* evidence to establish the causes of death of his patients and make diagnoses. He thus used his findings as evidence to understand individual diseases. However, disputes arose over the scientific status of anatomopathological correlations: it was extremely difficult to confidently establish which lesions were the result of disease and which were the result of thanatomorphosis. From 1666–1692, Malpighi and colleagues recorded their findings in notebooks that testify to their intensive and collaborative dissection practices. In 1675, Giovanni Battista Capponi (1620–1675) died and was dissected by his friend Malpighi. Beyond his role in the microscopic exploration of living organisms, Malpighi promoted the development of a mechanistic view of pathology.

Indeed, pathological anatomy played an important role in his commitment to mechanical anatomy (Bertoloni 1997), in which dissections, like the microscope, were able to reveal the structure and mechanical function of organs. Francesco Albertini (1662–1738), a student of Malpighi, wrote a text on respiratory diseases in 1726 on the basis of repeated and comparative dissections. He systematically explored the signs of diseases and the affected parts and conducted his research in the context of clinical medicine, defined by the correlation between diagnostic signs and specific lesions. He proposed a classification of respiratory diseases on the basis of the repetition of identical lesions and correlative clinical signs. He classified anatomical signs and their clinical consequences into two categories: aneurysmal defects with dilation of the left atrium of the entire heart or the aorta, which produced abnormally prolonged palpitations in the chest accompanied by vibrations and blows; and varicose defects, which are dilations of the inferior vena cava or the right atrium or right ventricle, the pulmonary vein or artery, causing unusual movements in the torso but without palpitations:

> Previously, we rarely, if ever, mentioned damaged structures of the heart and precordium in cases of impaired breathing. Nowadays, after numerous observations made on the dead, these terms are heard too often and are feared from the outset. (Albertini 1731, p. 404)

He focuses his studies on structural changes in parts of the body associated with specific manifestations of diseases observed in bedridden patients and has two complementary objectives: to understand normal structures and their pathological variations in the context of disease studies. Clinical anatomy provides a microscope for viewing pathological conditions.

Michele Lipari, known for his 1665 publication, *Galenistarum triumphus novatorum medicorum insanias funditus eradicans*, shared Malpighi's belief in the heuristic value of observations made on cadavers to identify the physical origin of diseases.

At the end of the 17th century, practical anatomy provided a disciplinary framework for pathological dissections, with the goal of improving medical practice in terms of diagnosis and treatment. The disputes between clinicians Giovanni Girolamo Sbaraglia (1641–1710) and Malpighi involved discussions, on the one hand, around Malpighi's empirical and direct medicine, which was based on anatomical investigation and *postmortem* dissection, and, on the other hand, around medicine, which focused on clinical observations and treatments.

Practical rules were suggested by Johann Conrad Peyer (1653–1712) in *Methodus historiarum anatomico-medicarum* (Peyer 1678). He describes anatomical and clinical cases organized around a four-step approach:

– examination of all symptoms and ailments suffered by a patient;

– description of damaged organs, observed *postmortem*;

– account for the patient's individual characteristics (age, gender and temperament);

– assessment of the causes and origin of the symptoms and the disease.

Between casuistry and peculiarities, scholars from the 17th century followed a medical approach that combined humoral theories, Galenism and the exploration of new organic regions.

1.3. From moving bodies to organs in pieces

In some respects, the Cartesian mechanism freed the Aristotelian representation of a body in pieces, which was a full representation and shown as a body in motion. In Vesalius' treatises, fragmented parts are placed alongside these depicted bodies.

However, once the body is conceived as a machine composed of detachable parts, these parts become elements that can be studied in isolation and repaired. This is how Descartes' mechanistic views on the bodily machine came to impact the history of anatomy and medicine.

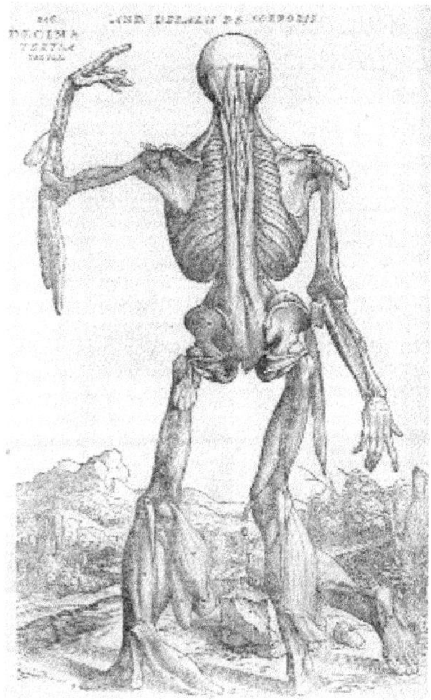

Figure 1.3. *An illustration from "De humani corporis fabrica" (Vesalius 1543)*

1.3.1. *The body: a whole broken down into its parts*

The concepts of the body and its parts have multiple relationships. Prior to the concept of the animal machine, Santorio Santori (1561–1636) designed a series of instruments intended to measure fundamental physiological processes, such as pulse and temperature, in isolation. He thus objectified physiology on the basis of a model in which each organ corresponds to a function that can be measured. For example, in 1602, he introduced the *pulsilogium*[8], which enabled him to quantitatively describe various regular and irregular pulse frequencies (Bigotti and Taylor 2017, p. 6).

During the 17th century, science took a Galilean turn, leaning toward founding knowledge on objective analysis of facts. Galileo Galilei (1564–1642) sought to apply the mechanical laws observed in the inorganic world to movements of the body (Galilei 1632). He drew analogies with mechanical weights, composed of systems of ropes and pulleys, to explain muscle movements. His approach influenced Cartesian studies, whose mechanism "authorizes" systematic

8 He also designed the thermoscope in 1611.

representation by organs. If the body is a machine, its parts can be represented separately for more detailed knowledge:

> [...] but also that it contains all the parts necessary to make it walk, eat, breathe, and finally imitate all those of our functions that can be imagined, so as to proceed from matter and only depend on the layout of the organs. (Descartes 1677, pp. 1–2)

This "machine" contains clocks, artificial fountains, mills and other mechanical assemblies that anatomical developments would strive to represent as nerves or tissues.

Figure 1.4. *Representation of human beings in motion obeying the laws of physics and mathematical rules (Descartes 1677, p. 76)*

In *The Passions of the Soul*, Descartes describes the body failure that causes loss of heat and movement, leading to a fatal outcome. No metaphysical cause is associated with the mechanical malfunction of the body:

> It consists in the fact that, seeing that all dead bodies are deprived of heat and, consequently, of movement, it was imagined that it was the absence of the soul that caused these movements and this heat to

> cease; it was thus believed, without reason, that our natural heat and all the movements of our bodies depend on the soul: whereas we should have thought, on the contrary, that the soul does not leave the body when we die, but only because this heat ceases and the organs used to move the body decay. (Descartes 1649, pp. 6–7)

The body becomes a machine that must be imagined both in terms of its dismantling and reassembly, an approach that lies at the foundations of physiological thinking. To construct an iatromechanics that is neither purely theoretical nor a priori, Borelli (1680) encouraged anatomy in *De Motu Animalium* to explore the smallest machines hidden within the organism. Thus, we can proceed with bodies in the same way as with a watch that is opened to examine its various parts (Gandolfo and Deschaux 2010, p. 177). He developed a quantitative approach to pathology and provided a dynamic foundation for Malpighi's exploration of living organisms. In the context of dissection, quantitative and mechanistic discourse favors the correlation of anatomic pathology with clinical findings. Indeed, if the "dissected" body loses the expression of its functions, then the correlation between intracorporeal lesions and clinical symptoms forms the framework for a possible understanding of diseases. Descartes saw health as the ultimate goal of anatomical knowledge (Cuir 2016, p. 213). The human body is an animalized machine, and "this entire Cartesian approach to the body influenced the Cartesian view of the role that medicine should play, as well as a new way of looking at disease" (Perru 2022, p. 8).

Antonio Maria Valsalva (1666–1723), Morgagni's teacher, established the application of detailed anatomy as one of his principles, which involves analyzing the smallest details to understand the functions of complex organs. He advocated the application of the concept of sympathetic physiology as an essential tool for understanding the coordinated functioning of anatomical parts, as well as a direct approach to normal structures and functions and the application of anatomical knowledge to clinical practice and surgery. His treatise on the human ear, *De Aure Humana tractatus* (Valsalva 1735), is devoted to the formulation of functional hypotheses on the basis of anatomy. He uses clinical practice to explore auditory functions. His medical practice is thus based on an anatomophysiological approach linked to anatomopathology. By analyzing anatomical structures in relation to their functions, he established the methodological necessity of applying reason for observation. This link was an essential element of his rhetoric of exploratory medicine, according to which medical practice is limited if it relies solely on observation without resorting to reason. The principle that disease can be conceived as an experience of nature and can shed light on structural and functional norms becomes predominant. From an epistemological perspective, the mechanistic approach is at the very core of exploratory medicine. By considering that biological tissues obey the laws of physics and are composed of matter and movements, bodies themselves are composed of tissues and, while being thought of in terms of their similarity to machines, refer to extremely complex, animated and living machines.

1.3.2. *The merging of mechanistic and vitalistic models*

By reducing the body to its elements, Cartesian theories expand on the exploration of a body determined by form and extension, with each perspective capable of being traced by a scalpel and mapped out. Western medicine developed as a way of thinking about space, and these explorations were supplemented by the notion of a system, which refers to a space conceived as the result of the totality of its properties (Malacarne 1798). These physiological systems, influenced by vitalist thinking, incorporate a mechanistic view that considers the particularities of a machine animated by internal organic movements and processes. An epistemological rupture – within which lies a thread, between mechanism and vitalism – occurs in the transition from a strict Cartesian mechanism to a conception of self-animated, self-preserving matter, whose functions are based on a set of organs, veins or arteries:

> The same can be said of the machines of our body, which are the basis of the machine: they are composed of strings, filaments, beams, levers, webs, flowing fluids, cisterns, channels, fires, sieves, and similar machines. By examining these parts through anatomy, philosophy, and mechanics, and also proceeding *a priori*, man has arrived at the formulation of models through which he demonstrates the causality of this effect and gives an *a priori* explanation for it; and through a series of the latter, aided by discourse, understanding the mode of operation of nature, he founded physiology and pathology, and consequently the art of medicine. (Malpighi 1697, p. 110)

The iatromechanistic program, according to which all vital acts are based on the intervention of mechanical forces and which aims to express physiology through mathematical formulas, has led to relatively few experimental achievements. Perceived as being too dogmatic and fairly ineffective in practice, it encouraged a revival of empirical medicine. A century later, the context of research into animal electricity by Luigi Galvani (1737–1798) (Galvani 1791) made it possible to explore living anatomy, understood in its vitalist and mechanistic foundations:

> Living, animated bodies are constantly electrified by the lungs, which are always in motion, and whose rest causes death. Air provides the lungs with electrical matter, as this element does in all other forms of electricity, by stripping itself of the fiery particles it contains. The arterial blood that comes from the lungs is the conductor of this electricity. The pulse of the left ventricle of the heart carries it throughout the brain through the expansion of the arteries, as in a Leyden jar, and from there it spreads to all the nerves. (Gautier-Dagoty 1775, p. 9)

From the 16th century to the 18th century, from mechanism to vitalism, medicine followed a dynamic that ranged from mathematical mapping to the study of how diseases manifest individually in the body. In *Préservatif contre la charlatanerie des faux médecins*, Guiseppe Gazola (1661–1715) (Gazola 1735) provides an account of the break between modern medicine and more traditional knowledge:

> [...] to successfully heal the Sick, one must not only have a perfect understanding of the location and appearance but also of the function of all the viscera of the human Body. One must understand the harmony of this Microcosm in order to deduce its disruptions and, through the symptoms and various effects of diseases, discover the various causes that produce them, as well as the activity of the remedies to be used to overcome them. Who can be better instructed in all this than a modern Physician? With the help of so many anatomical experiments and demonstrations of a Mechanics based on solid foundations, enlightened by experimental Philosophy, enriched by the effectiveness of chemical Remedies, and with the aid of Microscopes, which reveal the shape of the smallest parts that enter into the composition of mixtures, he can practice Medicine with dignity [...]. (Gazola 1735, pp. 190–191)

This excerpt describes the advent of a form of medicine that links ancient knowledge with a new understanding of animal physiology. Gazola highlights the importance of correlating knowledge of anatomy and physiology with the use of new instruments but also with topography and the concept of the "organic seat", the form and function of organs and their components, to determine the causes of pathological disorders. The mechanistic approach, which involves a fragmented view of the "pieces" of organic structures, allows for a transition from general anatomy to a breakdown by structures.

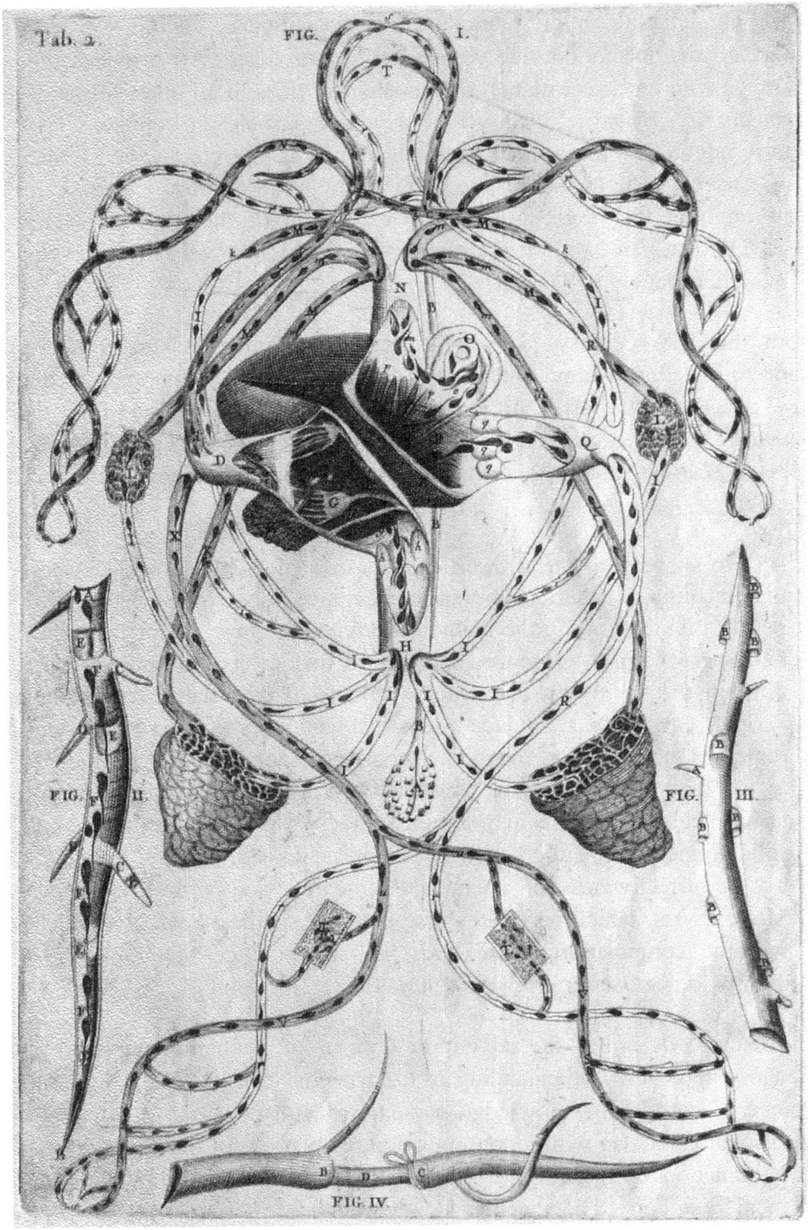

Figure 1.5. *Human machine represented through the automated lens by Reisel (1676, p. 16)*

In 1777, the *Journal des savants* published the "*Machine surprenante de l'Homme artificiel du sieur Reyselius*" (The amazing machine of the artificial man by Mr. Reyselius). This article presented Salomon Reisel's (1625–1701) machine, which imitated the circulatory system and followed the Cartesian philosophy of physiological movement made possible by organomechanical assemblies:

> This artificial man has vessels and viscera of the same form, structure, and size as those of a natural man, so that water or any other liquid poured into his mouth falls through a channel like the esophagus and soon reaches the right ventricle of the heart, after passing through a marvelous device in the intestines, the pancreas, the portal vein, and all the other areas that we have explained in our XII. Journal of this year. [...] One notices in this machine [...] all the other areas that are natural to man. (*Journal des savants* 1777, p. 252)

Medicine evolves through exploration, modeling and encounters or disruptions between different discourses. Being clinically oriented, it is complemented by rational medicine, which analyzes and seeks the organic causes of disease. The difficulties in defining its representations and rooting it in the human body, far from being merely formal, reflect profound conceptual changes in mechanistic and vitalist principles, tissue damage and imbalances between systems.

In his *Œuvres médico-philosophiques et pratiques* (Stahl 1863), Georg Ernst Stahl (1660–1734) broke down the concept of cause from a clinical and localization perspective of disease. He discusses occasional, predisposing, antecedent and continuous causes. A main cause corresponds, for example, to "[...] a certain substance, a certain mode, a certain disposition which, by its presence and especially by its immediate mode of action, prepares, provokes, and produces a morbid affection as its own direct effect" (ibid., p. 8). This categorization of causes enables dynamic thought on the sequence of physiological events that leads to pathology. He thus places the latter within a physiology that characterizes the temporality of the body; for example, vomiting can cause the expulsion of toxic substances and thus prevent a more serious morbid phenomenon (Stahl 1863, Vol. IV, p. 9).

From Malpighi to Valsalva to Morgagni, a conceptual, practical and medical line of thought converges, allowing links to be forged between anatomy, physiology, clinical practice and the causes of disease. These links revolve around the systematization of dissections since the 16th century but also around the evolution of intentions behind these practices: establishing the normal structures of the body, understanding its pathological states and finding the causes of death are all perspectives that enable medicine to think with and about the human body. Iatromechanism, which was developed from Descartes' concept of the "animal machine" to Julien Offray de La Mettrie's "man machine" (1709–1751), created the conceptual and medical environment in which Morgagni's work flourished, built around mechanism, materialism and vitalism.

1.4. Advent of the anatomopathological and clinical methods

Morgagni's research highlights a scientific encounter between Bonet's autopsy approach and Malpighi's exploratory objective. His work, *De sedibus et causis morborum per anatomem indagatis* (Morgagni 1761), was quickly translated into French (1765), English (1769) and German (1771) and is considered the birth of the anatomopathological and clinical methods. Within an extensive casuistry, the lesion is understood as the expression point of the clinical physiognomy of diseases (Keel 1984, p. 60). The way in which he follows the pathology in the organism reflects Valsalva's research and Malpighi's innovative spirit:

> If anatomy provides an advantage to good medicine by showing the origin and location of ailments, their causes and how they appear, from which we derive indications for choosing remedies, then the more accurate and precise it is; the more useful it will be. [...] Given that medicine, even practical medicine, does not consist in exposing remedies, but in considering signs, in the search for causes and indications to be deduced *a priori*: these things require an understanding of the economy of the animal, and consequently as accurate a knowledge as possible of the mechanical structure of the solid parts and the nature of the fluids that are processed in healthy intestines. (Malpighi 1697, p. 126, cited in Zampieri 2016, p. 225)

Morgagni's years in Bologna, between 1701 and 1706, were fundamental to the development of his method, which revolved around a complex in which anatomical location, physiopathological processes and clinical symptoms were closely related. His research was marked by the practical application of new ways of understanding morbid phenomena and by linking clinical manifestations with therapies. As a fervent advocate of the exploration of living organisms, Morgagni developed a heuristic intertwining of ancient and modern principles in which observation and empiricism were closely linked. He drew on clinical and pathological accounts from the work of Valsalva[9] and followed Borelli's iatrochemical methods, which he intertwined with iatrochemical principles, according to which diseases are understood as disturbances of the body's fluid systems. For Morgagni, diseases are entities rooted in the heart of organic structures and their processes. Although he did not use a microscope, he developed the exploratory aspect of diseases, understood as cross-dynamics between anatomy, clinical practice and pathophysiology.

1.4.1. *Anatomical explorations and location of diseases*

While Vesalius initiated a major break with traditional human anatomy, "[...] it was not until Morgagni that clinical symptoms and organic lesions began to be linked on an experimental basis, showing that every functional alteration that manifests itself symptomatically necessarily corresponds to an organic alteration" (Laplantine 1993, p. 57).

Morgagni was part of the Italian medical and biological culture that formed developments in anatomy. In 1706, he published *Gli adversaria anatomica prima*, a treatise in which he described a series of small anatomical parts, such as the glands of the trachea and the male urethra (Belloni 1980, p. 242). These descriptions represent new contributions to the mechanistic structure of the organism. This work led to his appointment to teach theoretical medicine in Padua (1711) at the age of just 24, on the basis of the texts of Avicenna's canon, Galen's medical arts and Hippocrates' aphorisms, before devoting himself to teaching the macroscopic pathological anatomy of organs and the groups they form. The laws of logic guide reflection and research when there are not enough facts. This, at least, is the path between rationalism and empiricism that Morgagni[10] followed in the treatment of diseases. The treatise, published in 1761, consists of five books, each devoted to a part of the body:

– book 1: diseases of the head;

– book 2: diseases of the torso;

9 For example, he recounts the medical history of a young man who died after a series of delusional episodes, published in Letter 7 entitled "On Phrenesia, Paraphrenia, and Delirium": "Upon opening the skull, a gelatinous concretion was noticed accompanying the sides of the blood vessels crawling through the pia mater. This meningeal membrane had torn at the base of the brain, and a quantity of fluid similar in color and consistency to cow's milk flowed out" (Morgagni 1820–1824, vol. 1, p. 390).

10 In 1761, he referred to the treatise published by Antoine Arnauld (1612–1694) and Pierre Nicole (1625–1695), *La logique ou l'art de penser* (Arnauld and Nicole 1662).

– book 3: diseases of the stomach;

– book 4: surgical and universal diseases (fever, tumors and ulcers);

– book 5: additions to the previous sections.

The 70 letters that make up these five books follow a careful order, on the basis of the scalpel cuttings, that considers a pathophysiological mechanism, on the basis of clinical signs that are evident in vivo and on organic lesions found in the corpse. Each of these letters begins with a nosological type of description and a complete list of clinical signs. He then discusses the different species, as well as the specific characteristics of the morbid variations observed, highlighting similarities, differences and possibilities for classification. He refers very regularly to the *Sepulchretum* (Bonet 1679). He also followed Sydenham's principle, according to which the lesion of an organ corresponds to the immediate cause of the disease, and postulated the use of inductive casuistry related to the compilation of observations. Morgagni thus objectified the idea that the immediate cause of disease is located in organs. Taking up themes that had been focused on in previous centuries around the causality of diseases, he developed the notion of efficient cause, which frees evils through the influence of habits. He thus compiled a range of possible factors that were likely to cause diseases: they were congenital or hereditary, seasonal or related to the age of the subjects. These descriptions were based on clinical biographies written with meticulous attention to detail. Morgagni refers to the seat of the disease as the place where it manifests itself, from a dynamic and causal perspective that involves a mechanistic aspect: if the body is a machine, it is made up of localized mechanisms located in specific places, on which its functions and links with other organ systems depend. The pathological model of the mechanism involves a disease that is located where health disruption occurs. The anatomopathologist provides a statistical approach to this research by comparing symptoms and lesions in individual cases as well as clinical aspects, with the goal of understanding the pathological phenomenon. They attempt to establish a series of stable relationships that repeat themselves beyond singularities and can contribute to the formation of universal knowledge. During the examination of the corpse of a man suffering from mania, they highlight the difficulties in determining the cause of the disease and identifying the lesions responsible for it. He noted that when the brain was opened, "black, liquid blood distended the vessels of both meninges [...]" and that "[...] nothing was more remarkable than the hardness of the brain" (Morgagni 1820–1824, p. 427). His causal descriptions fluctuate between exciting causes and predisposing causes. These two categories correspond to two uses of the concept of cause: clinical use and anatomopathological use.

Clinical use, related to the temporality of the disease, is expressed by Morgagni through the notion of predisposing causes and immediate causes, which are seen as the factors necessary for the expression of the disease. For example, he explains nosebleeds as the erosion of blood vessels, which represents a risk factor and will undoubtedly cause bleeding in the near future. In his descriptions, he intertwines iatromechanistic principles about the structures in which blood circulates and iatrochemical considerations about the aqueous nature of this fluid, which promotes hemorrhage and facilitates the erosion of vessels. During his anatomopathological explorations, he attempted to differentiate between distant causes, such as a lesion of the viscera that could trigger a hemorrhage, and immediate causes, such as a lesion of the nose vessel alone. Beyond the clinical aspect of his research, dissection proved to be the criterion of these considerations. One of his hypotheses was that, on the basis of visible lesions, it was possible to trace the course of the physiopathological events that preceded them. Starting from the lesion, Morgagni "imagined" what might have caused it, thus shedding light on physiological mechanisms through pathological anatomy. The middle ground of the Hippocratic heritage with Bonet's research, the contributions of mechanism, Galilean rigor and the organs' tether to the heart can be seen in the clinical descriptions, followed by the dissection report. He thus introduces:

> [...] a pathology based on a mechanistic anatomy, according to which organs are machines obeying the laws of Galilean and Newtonian physics. Thus, the causes of disease were reduced to the mechanical lesion of an organ, quantitatively determined or determinable, which could no longer perform its function. (Zampieri 2016, p. 61)

Morgagni detailed more than 700 dissections. Rejecting theoretical speculation[11], he followed in the footsteps of Galilean experimentalism, according to which the book of nature is the specific purpose of science:

> In the first, I refute the opinion of some physicians who have dared to raise doubts about its usefulness, and I show anatomists, who establish the seat and cause of diseases based on dissection, how to be careful and avoid easy mistakes that present themselves. [...] The third mainly serves as a response to those who claim that dissections are useless because they cannot reveal the primary causes of diseases,

11 He takes a stand, for example, against the Cartesian theory of the pineal gland as the seat of the soul: "We must no longer hope to confirm Descartes' opinion on the seat of the thinking soul, based on the integrity of this gland in individuals who are not insane, or based on the lesions of those who were affected by madness" (Premuda and Bonuzzi 1982, p. 77).

> which remain hidden and completely inaccessible to our senses; as if they did not serve to discover some that are internally obvious, and as if knowledge of them were useless because, even when known, there are a considerable number of diseases that cannot be cured. In the fourth, I seek to determine which is more useful: opening the body after a rare disease [...] or after a more common disease. (Morgagni 1824, vol. 1, p. 48)

Morgagni's study of corpses was crucial to the development of medicine. His solidist thinking[12] led him to recognize that alterations in the body's structure were not only indicative of disease, its location and its causes but also of factors disrupting organic functions and symptoms of various pathological processes. Morgagni starts from the concept of the organism as a machine, whose health results from the harmonious functioning of both the visible and the hidden smaller parts that compose it. Like machines created by humans, organic machines wear out and deteriorate, resulting in deterioration of their functioning. Disorders in animal health manifest themselves in clinical symptoms proportional to the location and nature of the damage caused. This is reflected in the title of his work, which emphasizes his intention to model a dynamic perspective of the interactions between disease and organic structures. The symptoms resulting from functional alterations in the latter. The parallelism between lesions and symptoms constitutes the anatomo-clinical histories that feed into his treatise. This discursive structure, in which biography meets clinical practice and is expressed at the core of the organs, produced the epistemological structure of a new form of knowledge: anatomical pathology and clinical knowledge.

In the letter LXI, entitled "Delusions that occur without fever", we can distinguish five stages in his approach:

– Start by setting out the general framework of the study, which focuses on dementia without fever, within which he discusses three dissections.

– Within the general framework of dementia without fever, distinguish the specific case of a woman with dementia, providing her known history, age, and physical condition.

– Move on to the examination of the corpse:

> Upon opening the skull, I noticed only one thing about the meninges, namely that when I barely grasped and touched the pia mater, it

12 The solidist model attributes all diseases to a change in the vital properties of organs considered as solids.

> moved along and was very easily removed from the crevices of the brain. There was therefore no doubt that there had been water beneath it, and a small quantity was even found in the ventricles; [...] and yet the substance of the brain and cerebellum was harder than usual. (Morgagni 1824, p. 497)

– Formulate a general law on the basis of the correlation between clinical and anatomopathological elements:

> [...] note, among other things, that the small stones he found in the pineal gland were also yellow or yellowish, and then see what other lesions, which I will soon discuss [...] not only in this gland but also in the brains of the insane [...]. (ibid., p. 502)

– Assess one's discourse with empirical perspectives:

> Thus, convinced that further research is needed on this subject, I will move on to what has occurred to me in the brains of the insane, if not always, then at least almost always [...]. (ibid., p. 505)

Morgagni taught how to deduce alterations from the form of local lesions and, conversely, how to identify chemical alterations from visible lesions and mechanically trace them back to functional alterations. This interpretation of the body and its mechanisms refers to the polymorphism of clinical phenomena. Organic lesions are decisive for the clinical form of diseases, provided that they can be identified as the cause. Once identified, their location is integrated into a dynamic conception of the organism. Diseases are therefore the result of biological experience and disruptions at the core of physiological mechanisms. In the letter LXII, "On epilepsy, convulsions, and paralysis", Morgagni considers disease through the prism of two peculiarities: that of the clinical form and that of the organism affected by it. Each disease is therefore specific through the combination of symptoms and the organic field that expresses them. He moved from theoretical medicine to practical medicine. His descriptive rigor led to significant contributions in the most diverse chapters of medical science. His method was the result of developments in the debates between neoteric medicine, empirical medicine and ancient medicine, which was revived in the form of neo-Hippocratic medicine. His treatise became crucial for clinicians, as knowledge of the anatomical site of a disease could guide treatment; for pathologists, knowledge of the clinical symptoms associated with organic lesions could guide diagnosis; and for surgeons, knowledge of the diagnosis and the corresponding anatomical site could guide any intervention aimed at healing. Morgagni thus shifted the clinical focus to internal structures. By linking practical anatomy and empirical medicine, he followed a path that led from the opening of corpses to the study of lesions in organs without resorting to the microscope, preferring to use the anatomical scale of the organ instead of the latter. While this may be seen as technical skepticism, there are also epistemological reasons for this attitude: Morgagni refused to search for ultimate causes in pathological processes, considering the former to be obscured and inaccessible. Although the microscope made it possible to explore *infinitely small* things, it was not possible to clearly identify what was shown, causing numerous speculations. Morgagni can be considered a pioneer of the Paris School of Anatomy. His method was adopted in France by Théophile-Hyacinthe Laënnec (1781–1826) and perfected by Bichat. Therefore, medicine had to seek out internal organic lesions so that each disease could be understood and classified through a clinical reading of the body.

1.4.2. *The significance of dissections to renewed surgical practice*

The systematization of dissections has made it possible to establish practical training in *postmortem* surgery. From the 17th century to the 18th century, the dissection table became the place where anatomists would open bodies. While they had a medical goal, their learning involved exploring pathological processes, which were seen as anatomical transgressions. These processes shed light on the anatomical corpus while also leading to the discovery of new therapies. The pathological body thus occupies several areas of analysis: normal anatomy, clinical correlations, and research into the causes and effects of disease. It is also linked to the rethinking of surgical practice. The anatomical pathology collections of Frederik Ruysch (1638–1731) are used in the training of surgeons to illustrate the different categories of operations: reductive, extractive, etc. The evolution of dissection techniques intersects with concerns about the preservation of the human body.

In the 17th century, Jan Swammerdam (1637–1680) had the ingenious idea of using a hot substance that would cool down to reveal the tree-like structure of blood vessels. From 1650 onward, Dutch scholars injected mixtures of spirit of wine or turpentine essence to give substance to anatomical illustrations. The embedding of this stage completed the scientific groundwork for the objectification of the body and its mapping. The mapped organ systems became topographical landmarks that surgical students could use to learn how to operate. In this context, Ruysch focused his work on surgical anatomy (Ruysch 1691, 1737). He approached pathologies from the perspective of operations that were not classified by the anatomical or surgical field, indicating the overlapping boundaries between the two fields. Knowledge and skills in dissection were fundamental to both, enabling the transition from one to the other.

Herman Boerhaave (1668–1738) defined pathology as "the second branch of physics [...] dealing with diseases, their differences, their causes and their effects, or symptoms by which the human body is known to vary from its state of health" (Boerhaave 1742–1746, Vol. 1, p. 77). While physiology questions the body about its natural mechanisms, pathology questions it in the context of disruption of the animal economy. While Boerhaave asked theoretical questions, Ruysch took a practical approach, focusing on the potential improvement of surgical practices. However, surgery refers to a localizing practice that requires delving into the depths of anatomy. Anatomy is similar to surgery: by being objectified, the cause of the lesion becomes an operable reality. By objectifying disease through lesions, anatomical pathology gives substance to pathological processes, showing their embodiment and enabling their extraction or reduction to be conceived. The cartographic approach, which is based on Renaissance illustrations, is consistent with surgery focused on therapy.

Vincenzo Malacarne (1744–1816), the successor to Morgagni's anatomopathological and clinical approach, developed ideas related to the changes needed in the art of surgery. Surgeons need to adopt a more medical approach and acquire solid knowledge of anatomy to restore pathological conditions to normal health. In 1794, he published a manual on traumatic anatomy called *Ricordi d'anatomia traumatica pubblicati ad uso dei giovani chirurghi militari di terra e di marina* (Remarks on traumatic anatomy published for the use of young military surgeons on land and at sea) in which he proposed surgery to be divided into categories: reduction, separation, reattaching, extraction, amputation, replacement and addition. The extraction category is an important surgical category, as the care of patients with hydrocephalus or cretinism, as well as cancer, depends on its development. The adaptation of surgical operations to the particularities of each organic structure is recommended:

> With extraction, we free the patient from excess blood, milk, urine, feces, cataracts, and stones; we free them from diseases caused by foreign bodies that have deeply implanted themselves into the substance of the limbs, blocked up channels, and precipitated into the cavities. (Malacarne 1794, p. 213)

A practical surgical program was thus initiated between the 16th and 18th centuries. It accompanied and correlated with developments in anatomy, illustrating its therapeutic objectives. Anatomical observations, illustrations and preserved bodies or body parts played an educational role for the student surgeons, who had to practice on corpses and acquire perfect anatomical knowledge before they were able to operate. In addition, it was also necessary for them to have knowledge of the specific topography of each pathological process to be aware of the changes undergone by the organism in the event of a pathological condition. Surgeons must also operate on the regulation of fluids and substances in the organic environment, ensuring that the ideal balance between the different parts of the body is maintained, thus intervening in the physiological relationships that exist between the systems of the animal economy. Surgical techniques must be developed to be able to intervene in the regulation of the systems that make up the animal economy. Thus, from anatomical explorations to studies of disease sites, the prospect of a medicine capable of intervening in pathological processes and restoring the subject's health has emerged. Classical teachings were combined with the development of dissective and surgical techniques; the revival of a neo-Hippocratic clinical approach; and mechanistic, vitalist and materialist perspectives. Medical practice has been shaped by this intertwining of medical traditions and scholarly discourse.

1.5. Schools of thought modeling medical practice

Vesalius can be seen as having liberated human anatomy from Galenic authority. By prioritizing direct observation, the Renaissance anatomist contributed to the emergence of an innovative form of medicine: "Naturally, proponents of neoteric medicine could only call for direct observation of nature, since it was specifically new anatomical discoveries that served as the basis for iatromechanical interpretations" (Zampieri 2016, p. 55).

Thus, two branches of medicine have been developed: neoteric medicine, represented in particular by micrologists, and empirical medicine. The former sought to observe the microstructures of the organism and innovate in the fields of protocols and exploration techniques, whereas the latter focused on the external manifestations of diseases by developing nosology. Clinical reading was combined with anatomical and pathological explorations and embodied in a medical practice of correlations on the basis of the teaching of the senses, guided by experience.

While the 17th century saw mechanism favoring the development of anatomy, in the 18th century, pathology became an anatomical discipline known as "medical anatomy" or "practical anatomy", giving rise to pathological anatomy, which was then fully developed in the 19th century. Although Morgagni's pathological anatomy was based on the potential nature of readability and consistency, the conjectural aspect of rational deductions could not be dismissed by scholars, who found themselves obliged to combine empiricism, rationalism and mechanism. By meeting these principles, disease became an objectifiable entity on the basis of the relationships among the anatomical site, physiopathological processes and clinical symptoms, regardless of patient diversity. Exploratory, pathological and clinical medicine resulted from various theoretical legacies that marked certain ambiguities. One of the challenges that both Sydenham and Morgagni faced was establishing a science of disease while considering the unique characteristics of each case. To do this, the clinical eye and the anatomist's hand worked together to form what could be considered a microscope of the body, capable of detecting sites of imbalances and recognizing their external signs. Physicians have correlated the classical knowledge of humoral medicine with medical empiricism and an exploratory approach.

In the 17th century, the human body became mechanized while simultaneously being rationalized through Galileo's perception of a nature that could be accessible and transcribed into scientific laws. Experimental philosophy, which is based on geometric principles and physical, mechanical and chemical experiments, paved the way for medical science. This encounter between rational mechanism and the direct observation of anatomists formed the virtual microscope that supported Morgagni's writings. It resulted in a multifaceted interpretation of the human body, viewed both in terms of the universality of its structures and systems and the uniqueness of its pathological expression. Anatomy thus permanently became part of clinical practice and surgery, providing numerous medical perspectives.

The 17th century was a period of great medical systems, which included the establishment of a complex medical interpretation framework. Anatomists, physicians and surgeons, claiming to follow Hippocrates, Sydenham or Bacon, were, in turn, supporters of observing bedridden patients, accurately recording symptoms, using a mechanistic approach to understand how the body works when healthy and when sick, experimentation and anatomical investigations. Medical animism was revived in the work of Stahl, who emphasized the uncertainty of medicine and urged scholars to adopt caution in the same way that a watchmaker should when opening a closed watch. While he described the mechanistic reduction of functions, he attributed the regulation of vital movements to the soul. If the body is a machine, it needs additional movement reintroduced through the prism of the soul.

This animistic perspective coincides with the mechanism of Friedrich Hoffmann (1660–1742), according to whom mechanism and physics are essential to medicine. A proponent of an iatrophysical approach, Hoffman saw himself as a physician-geometer. While claiming to conform to the known laws of mechanics, he remained within the limits of idealized mechanics to account for the particularities of organ movements. The clinic, which was a successor to the biographical aspect developed by the Hippocratic school, adopted an approach similar to that of the phytographer from Sydenham and continued through the work of François Boissier de Sauvages de Lacroix (1706–1767).

The latter, whose brother was a botanist, published the *Nouvelles classes des maladies* (New classes of diseases) in 1731, which included the genera and species of all diseases, with their signs and symptoms, in an order similar to that used by botanists. He developed a program for studying nosological species and varieties, adopting an impactful taxonomic approach from Carl von Linné (1707–1778). By directing medical attention toward the concept of clinical tables, he refrained from speculating about supposed functional disorders. Medicine was at the center of many systems and in the midst of intense technological development. Theoretical anatomy was based on healthy bodies, practical anatomy was based on people who had died of unknown diseases, and the two came together in the practice of autopsy. Morgagni rooted his theoretical ideas in the rational medicine of the 17th century, in a clinical approach, but also in Vesalius' theoretical anatomy. Following the Paduan physician, "we can therefore define disease as a vicious disposition, an impediment of the body or one of its organs, which causes a more or less noticeable lesion in the exercise of one or more functions of healthy life, or even causes one or all of them to cease completely, except for the movement of the heart" (Jaucourt 1765, p. 929).

Between the end of the 18th century and the beginning of the 19th century, a clinical view of disease swept across Europe, which was deeply rooted in pathological anatomy and developments in hospital structures.

1.6. Developments in clinical anatamopathology

At the end of the 18th century, a new approach to the clinical examination of the body complemented Morgagni's methods. This was the technique of auscultation, described in 1761 by Léopold Auenbrugger (1722–1809) in his treatise *Inventum Novum ex percussione thoracis humani ut signo abstrusos interni pectoris morbos detegendi*. In the Preface, the scholar announced that his approach "consists of the percussion of the human thorax, whose various resonance of sounds determines the judgment on the internal state of this cavity" (Auenbrugger 1855, Preface).

In addition, Jean-Baptiste Bouillaud (1796–1881), one of the founders of the Parisian clinical school, helped to systematize the principles that were inherited from Morgagni's anatomo-clinical method. The scientific environment favored the application of these principles on a larger global scale, owing in particular to the establishment of hospitals dedicated to medicine. Hearing is added to sight and touch to listen to the sounds of the body and its internal movements, thus providing a sensory basis for physiology, whose mechanisms are disrupted in the *postmortem* context. Although percussion was used as a diagnostic tool before 1800 by Heinrich Callisen (1740–1824) in *System der Wundarzneikunst* (1788) or in France by

Raphael Bienvenu Sabatier (1732–1811) in his *Traité d'anatomie complet* (Sabatier 1791), it appears that the medical community did not take to it immediately. The scholar's method remained confined to local teaching at several German universities, such as those in Halle, Wittenberg and Würzburg. Among the reasons that can be put forward, we can hypothesize that it was necessary to wait for the widespread dissemination of Morgagni's approach in Europe. This reached France in the early 19th century, at a time when scholars were becoming fascinated by this profound auditory reading of the diseased body, prior to dissection. The Italian method of anatomopathological investigation and the Austrian method of clinical examination were combined in Paris. A French translation of the Austrian text, *La nouvelle invention*, was published in 1808 by Jean-Nicolas Corvisart (1755–1861), who paid tribute to the method devised by Auenbrugger by rescuing it "from the total and extraordinary oblivion into which it seemed, to me, to have fallen in France in particular" (Corvisart 1855, p. 183).

In addition to an interpretation based on the quasimusical rhythm of the body, Auenbrugger depicted its functioning in vivo. In 1767, Joseph Lieutaud (1703–1780) published *Historia anatomica Medica*, a treatise in which he described 3,500 clinical cases accompanied by anatomopathological reports, thus highlighting France's reception of Morgagni's research.

In Leiden, Eduard Sandifort (1742–1814) published an illustrated work in four volumes entitled *Observationes anatomico-pathologicae* (1777–1781), whereas in London, John Hunter's treatises on pathological anatomy appeared, in which he provided precise observations, thus countering the subjective aspects of accounts gathered from patients. He developed an experimental study of pathological processes to understand the dynamics of lesions (Hunter 1786). In 1793, Matthew Baillie (1761–1823) published *The Morbid Anatomy of Some of the Most Important Parts of the Human Body*, in which he emphasized the importance of a dynamic approach to pathology. Aware of the margin of uncertainty and possible errors between the lesion and the dysfunction that appears, he focused on the study of abnormal structures, with the aim of revealing the smallest parts of the body where morbid actions occur.

During the 18th century, general, practical and histological anatomy were added to microscopic anatomy. Clinical anatomy complemented these approaches, which differed in terms of their purpose and the scale of their studies. Bichat's histology introduced an additional scale of exploration between microscopic organisms and organs. He developed the idea that each tissue is susceptible to disease independently of whole organs, notably, in 1799, in *Le traité des membranes en général et de diverses membranes en particulier* (Treatise on the membranes in general, and on different membranes in particular). In 1801, he stated the following:

> The more we observe diseases, the more we become convinced of the need to consider local diseases, not in terms of the composite organs they almost never affect in their entirety, but in terms of the various tissues they almost always attack in isolation. (Bichat 1834, p. 27)

The tissue became the site of normal and pathological processes, as well as support for the lesion. Laënnec then systematized the use of the stethoscope, which was invented in 1819. For him, the local lesion was neither the cause nor the effect of the disease, but rather its essence. The clinic needed to conduct a study of the living body to find internal lesions. His work represents a synthesis of the classes he took in 1801 with Corvisart and those of Guillaume Dupuytren (1777–1835), with whom he quarreled over the links between the internalization and externalization of pathologies.

Given these controversies, Laënnec promoted a more anatomical approach to medicine and a more medical approach to surgery. In 1819, in *Traité de l'auscultation médiate*, he summarized anatomy and clinical practice, thereby intertwining medical and surgical thinking. Rejecting strict organicism, he focused on the dynamics of lesions, which he did not conceive according to their location but rather according to their form. Thus, at the end of the 18th century, concepts that were directly inherited from Morgagni, moving from the organ to the tissue, developed. This anatomo-clinical movement was accompanied by a renewal of outpatient care, which was based on new tools and a new way of listening to the body's functions, laying the foundations for the Parisian clinical school.

1.7. Conclusion

Between the 16th and 18th centuries, anatomy developed and gradually specialized in etiology, systematically seeking links between alterations in organic structures and signs of disease:

> As for anatomopathology, although it uses a different classification system from the previous one (based on lesions visible in organs (Morgagni) or tissues (Bichat)), it differs even more from specific medicine, because the location of the cause is considered to be integral in determining the "seat", and the disease completely coincides with its bodily manifestation; it is nothing else. (Laplantine 1992, p. 58)

Thus, while anatomy was the dominant trend in Western medicine in the 16th century, it became mechanistic from the 17th century onward, with the electrification of corpses occurring during the 18th century. The organism, understood in dynamic and functional terms, became a place where it was possible to explore many avenues for an approach characterized by anatomical and pathological thinking.

1.8. References

Primary sources

d'Acquapendente, G.F. (2013). *De venarum ostiolis*. Literary Licensing, LLC, Whitefish.

Albertini, I.F. (1731). *Animadversiones super quibusdam difficilis respirationis vitiis a laesa cordis et praecordiorum structura pendentibus*. In *De Bononiensi Scientiarum et Artium Instituto Atque Academia Commentarii*, volume I. Bologna, 382–404.

Arnauld, A. and Nicole, P. (1662). *La logique ou l'art de penser*. Tel Gallimard, Paris.

Auenbrugger, L. (1761). *Inventum novum ex percussione thoracis humani ut signo abstrusos interni pectoris morbos detegendi*. Joannis Thomas Trattner, Vindobonae.

Auenbrugger, L. (1808). *Nouvelle méthode pour reconnaître les maladies internes de la poitrine par la percussion de cette cavité*, J.-N. Corvisart (trans.). Migneret, Paris.

Auenbrugger, L. (1855). *Nouvelle méthode pour reconnaitre les maladies internes de la poitrine*. Adolphe Delahays, Paris.

Baillie, M. (1793). *The Morbid Anatomy of Some of the Most Important Parts of the Human Body*. J. Johnson, London.

Bartholin, T. (1653). *Vasa lymphatica, nuper Hafniae in animantibus inventa, et hepatis exsequiae*. G. Holst, Copenhagen.

Bayle, G.-L. (1855). *Recherches sur la phtisie pulmonaire*. Adolphe Delahays, Paris.

Beniviani, A. (1507). *De abditis morborum Causis*. Filippo Giunta, Florence.

Berengario da Carpi, J. (1523). *Isagogae breues, perlucidae ac uberrimae, in anatomiam humani corporis a communi medicorum academia usitatam*. Benedictus Hector, Bologna.

Bichat, X. (1799). *Le traité des membranes en général et de diverses membranes en particulier*. Veuve Richard, Paris.

Bichat, X. (1801). *Anatomie générale appliquée à la physiologie et à la médecine*. Brosson, Orléans.

Bichat, X. (1834). Anatomie générale appliquée à la physiologie et à la médecine. In *L'encyclopédie des sciences médicales*, volume 3, Allibert, M. (ed.). Virey, Paris.

Boissier de Sauvages de Lacroix, F. (1731). *Nouvelles classes des maladies qui, dans un ordre semblable à celui des botanistes, comprennent les genres et les espèces de toutes les maladies, avec leurs signes et leurs indications.* B. d'Allanville, Avignon.

Bonet, T. (1679). *Sepulchretum sive Anatomia practica, ex cadaveribus morbo denatis, proponens historias et observationes omnium pène humani corporis affectuum.* L. Chouet, Geneva.

Borelli, G.A. (1680). *De motu animalium.* Angeli Bernabo, Rome.

Borelli, G.A. (1989). *De motu Animalium (On the Movements of Animals)*, P. Maquet (trans.). Spinger Verlag, Berlin/New York.

Callisen, H. (1788). *System der Wundarzneikunst.* Proft & Storch, Copenhagen/Leipzig.

Coiter, V. (1566). *De ossibus et cartilaginibus corporis humani tabulae.* Joannem Rossius, Bologna.

Corvisart, J.-N. (1855). *Essai sur les maladies et les lésions organiques du cœur et des gros vaisseaux.* Adolphe Delahays, Paris.

Da Carpi, B. (1529). *Isagogae breves, plucide ac uberrime in anatomia humani corporis a communi medicorum academia usitata.* Benedict Hector, Paris.

Descartes, R. (1649). *Les passions de l'âme.* Henri Le Gras, Paris.

Descartes, R. (1677). *L'Homme de René Descartes et un traité de la formation du fœtus du même auteur, avec les remarques de Louis de la Forge*, 2nd edition. Théodore Girard, Paris.

Dodoens, R. (1581). *Medicinalium observationum exempla rara.* Cholinum, Maternum/The Wellcome Library, London.

Estienne, C. (1545). *De dissectione partium corporis humani libri tres.* Simon de Colines, Paris.

Falloppia, G. (1561). *Observationes anatomicae.* Marco Antonio Ulmo and Gratioso Perchachino, Venice.

Falloppia, G. (1563). *Libelli duo, alter de ulceribus, alter de tumoribus praeter naturam.* D. Bertellus, Venice.

Fernel, J. (1542). *De naturali parte medicinae.* Simon de Colines/Adam Saulnier, Paris.

Fernel, J. (1567). *Universa medicina.* Andreas Wechel, Paris.

Fernel, J. (1655). *Les sept livres de la thérapeutique universelle de Jean Fernel.* Jean Guignard père et fils, Paris.

Galeni, C. (1990). *De locis affectis libri sex.* Omnisys, Cambridge.

Galilei, G. (1632). *Dialogo sopra i due massimi sistemi del mondo Fiorenza*. G.B. Landini, Rome.

Galvani, L. (1791). *De viribus electricitatis in motu musculari commentarius*. Instituti Scientiarum, Bologna.

Gandolfo, G. and Deschaux, O. (2010). Histoire de la découverte du cerveau et de l'évolution des méthodes d'exploration. *Biologie-Géologie*, 4, 171–192.

Gautier-Dagoty, J.-F. (1775). *Exposition anatomique des organes des sens, jointe à la névrologie entière du corps humain, et conjectures sur l'électricité animale et le siège de l'âme*. Demonville, Paris.

Gazola, G. (1735). *Préservatif contre la charlatanerie des faux médecins*. Jean-Luzac, Leiden.

Harvey, W. (1628). *Exercitatio anatomica de motu cordis et sanguinis in animalibus*. G. Fitzer, Frankfurt.

Hoffmann, F. (1740). *Opera omnia physico-medica*, 6 volumes. de Tournes, Geneva.

Hunter, J. (1786). *Observations on Certain Parts of the Animal Oeconomy*. Cazin, London.

de Jaucourt, L. (1765). Maladie. In *Encyclopédie, ou Dictionnaire raisonné des sciences, des arts et des métiers*, volume IX. Société typographique, Neuchâtel.

La Mettrie, J.O. (1748). *L'homme machine*. Ellie Luzac, Leiden.

Laennec, R.-T.-H. (1819). *Traité de l'auscultation médiate*. J.-A. Brosson and J.-S. Chaude, Paris.

Laennec, R.-T.-H. (1884). *Traité inédit sur l'anatomie pathologique, ou Exposition des altérations visibles qu'éprouve le corps humain dans l'état de maladie*. Felix Alcan, Paris.

Laplantine, F. (1992). *Anthropologie de la maladie : étude ethnologique des systèmes de représentations étiologiques et thérapeutiques dans la société occidentale contemporaine*. Payot, Paris.

Lieutaud, J. (1767). *Historia anatomica medica*. Apud Vincent, Paris.

Malacarne, V. (1794). *Ricordi d'anatomia traumatica, pubblicati ad uso de' giovani chirurghi militari di terra e di marina*. G.A. Pezzana, Venice.

Malacarne, V. (1798). Dell'esistenza e dell'influenza di diversi sistemi nell'economia animale; Divisione generale de'sistemi, e meravigliosa estensione del sistema cutaneo. In *I commentari medici*, volume 2, Brera, L.V. (ed.). Callazzi, Pavia.

Malpighi, M. (1666). *De viscerum structura: exercitatio anatomica*. Iacobi Montij, London.

Malpighi, M. (1697). *Opera Posthuma. Figuris Æneis Illustrata*. A. & J. Churchill, London.

Marziani, P. (1626). *Magnus Hippocrates notationibus explicatus*. G. Mascardi, Rome.

Morgagni, G.B. (1706). *Adversaria anatomica prima ab eo nuper in eâdem academiâ public lecta, multis deinde accessionibus adaucta*. Typis Fernandini Pisarri, Bologna.

Morgagni, G.B. (1761). *De sedibus et causis morborum*, 2 volumes. Lutetiae, Venice.

Morgagni, G.B. (1820–1824). *Recherches anatomiques sur le siège et les causes des maladies*, 10 volumes. Caille et Ravier, Paris.

Paré, A. (1561). *Anatomie universelle du corps humain*. Jehan Le Royer, Paris.

Peyer, J.K. (1678). *Methodus historiarum anatomico-medicarum exemplo ascitis, vitalium organorum vitio, ex pericardii coalitu cum corde nati illustrata*. Apud Lambertum Roullaud, Paris.

Reisel, S. (1676). De statua humana circulatoria. In *Miscellanea curiosa medico-physica sive Ephemeridum Medico-Physicarum Germanicarum Academiae Naturae Curiosorum*, I(4/5), 1673–1674, Obs. CXCI, 245–246.

Rinaldi, M. (2018). Conjecture, signa et apparence. In *Hermes medicus. Discours et actions de l'herméneutique médicale (XV–XVII siècles)*, Carlino, A. (ed). Rodopi, Amsterdam.

Riolan (the Younger), J. (1618). *Anthropographia ex propriis et novis observationibus collecta, concinnata. In qua facilis ac fidelis... manuductio ad anatomem traditur*. H. Périer, Paris.

Riolan (the Younger), J. (1653). *Manuel anatomique et pathologique, ou Abrégé de toute l'anatomie, et des usages que l'on en peut tirer pour la connoissance, et pour la güerison des maladies*. Gaspar Meturas, Paris.

Ruysch, F. (1691). *Observationum anatomico-chirurgicarum centuria*. Henricum & Viduam Theodori Boom, Amsterdam.

Ruysch, F. (1700). *Epistola anatomica, problematica tertia et decima*. Wolters, Amsterdam.

Ruysch, F. (1737). *Opera omnia anatomico-medico – chirurgica*. Jansson Waesberge, Amsterdam.

Sabatier, R.B. (1791). *Traité d'anatomie complet, ou description de toutes les parties du corps humain*. Barrois, Paris.

Sandifort, E. (1777–1781). *Observationes anatomico-pathologicae*, 4 volumes. P.v.D. Eyk and D. Vygh, Leiden.

Santori, S. (1624). *Commentaria in primam Fen primi libri Canonis Avicennae*. Giacomo Sarcina, Venice.

Sassonia, E. (1607). *Opera practica, quibus hac novissima editione praeter alia accesserunt eiusdem auctoris tractatus De morbis mulierum & De pulsibus, ac urinis. Omnia quàm antè cura emendatiore*. Franciscum Bolzettam, Padua.

van de Spiegel, A. (1627). *De Humani corporis fabrica librii decem*. Tabulis XCIII, Venice.

Stahl, G.E. (1863). *Œuvres médicophilosophiques et pratiques de G.-E. Stahl*, 2nd edition, T. Blondin (trans.). Baillière, Paris.

Sydenham, T. (1683). *Tractatus de podagra et hydrope*. R.N. for Walter Kettilby, London.

Sydenham, T. (1882). *Le Traité de la goutte de Sydenham*, C. Lasègue (trans.). Asselin et Cie, Paris.

Sydenham, T. (1991). Thomas Sydenham's observationes Medicae (London 1676) and his medical observations. In *Manuscript 572 of the Royal College of Physicians of London*, Meynell G.G. (ed.). Winterdown Books, Folkstone.

Valsalva, A.M. (1735). *De aure humana tractatus*. Gisbertum Langerak, Paris.

Vesalius, A. (1543). *De humani corporis fabrica* [Online]. Available at: parisdescartes.fr.

Secondary sources

Addison, J. (2018). *Arts de la dissection*. Marguerite Waknine, Paris.

Blazy, M. (2016). *Morgagni. Le père de la médecine moderne*. CreateSpace Independent Publishing Platform.

Belloni, L. (1980). *Per la storia della medicina*. Arnaldo Forni Editore, Bologna.

Bertoloni, D. (1997). *Marcello Malpighi: Anatomist and Physician*. Leo S. Olschki, Florence.

Bigotti, F. and Taylor, D. (2017). Le Pulsilogium de Santorio : nouvelle lumière sur la technologie et la mesure au début de la médecine moderne. *Soc Politica*, 11(2), 53–113.

de Ceglia, F.P. (2020). *The Body of Evidence Corpses and Proofs in Early Modern European Medicine*. Brill, Leiden.

Cuir, R. (2016). *Renaissance de l'anatomie*. Hermann, Paris.

Daremberg, C. (1870). *Histoire des sciences médicales comprenant l'anatomie, la physiologie, la médecine, la chirurgie et les doctrines de pathologie générales*, volume 2. Baillère, Paris.

De Renzi, S., Bresadola, M., Conforti, M. (2019). *Pathology in Practice*. Routledge, London.

Grmek, M.D. (1990). *La première révolution biologique : réflexions sur la physiologie et la médecine du XVIIe siècle*. Payot, Paris.

Grmek, M.D. (1995–1999). *Histoire de la pensée médicale en Occident*, 3 volumes. Le Seuil, Paris.

Keel, O. (1984). La place et la fonction des modèles étrangers dans la constitution de la problématique hospitalière de l'école de Paris. *History and Philosophy of the Life Sciences*, 6(1), 41–73.

Klestinec, C. (2004). A history of anatomy theaters in sixteenth-century Padua. *J. Hist. Med. Allied. Sci.*, 59(3), 375–412.

Klestinec, C. (2010). Practical experience in anatomy. In *The Body as Object and Instrument of Knowledge*, Wolfe, C.T. and Gal, O. (eds). Springer, Dordrecht.

Ladislao, M. (1966). *Anatomica sive in cadaveribus sectis observationes, Marcello Malpighi.* Azzoguidi, Bologna.

Laplantine, F. (1993). *Anthropologie de la maladie.* Payot, Paris.

Mandressi, R. (2003). *Le regard de l'anatomiste.* Le Seuil, Paris.

Mandressi, R. (2013). Le corps des savants. Sciences, histoire, performance. *Communications*, 1(92), 51–65.

Montoliu, D. (2021). La médecine entre académie et université en Sicile au XVIIe siècle. Les *Dossiers du Grihl*, 14(2) [Online]. Available at: http://journals.openedition.org/dossiersgrihl/8744.

Ongaro, G. (2002). L'anatomia patologica e la clinica. In *Storia della scienza*, volume VI, Petruccioli, S. (ed.). Istituto della Enciclopedia italiana Treccani, Rome.

Perru, O. (2022). Descartes et Gassendi : la maladie est-elle naturelle ? *Bulletin d'histoire et d'épistémologie des sciences de la vie*, 1(29), 7–23.

Premuda, L. and Bonuzzi, L. (1982). Morgagni e la neuropatologia. *Acta Med Hist Patav*, 29, 75–81.

Roger, J. (2014). *Les sciences de la vie dans la pensée française au XVIIIe siècle : la génération des animaux, de Descartes à l'Encyclopédie.* Albin Michel, Paris.

Soulier, D. (2021). D'un Studium à l'autre : la peregrinatio academica des professeurs anatomistes en Europe (années 1540-1550). *Cahiers de la Méditerranée*, 102, 175–191.

Thomson, A. (2010). Les animaux plus que machines. *Dix-huitième siècle*, 1(42), 339–351.

Zampieri, F. (2013). The method of "rational medicine" of Giovanni Battista Morgagni. *Physis. Rivista Internazionale di Storia della Scienza*, 49(1–2), 85–127.

Zampieri, F. (2016). *Il metodo anatomo-clinico fra meccanicismo ed empirismo : Marcello Malpighi, Antonio Maria Valsalva e Giovanni Battista Morgagni.* 'L'Erma' di Bretschneider, Rome.

2

The Clinic or the Art of Diagnosis

Jean François THURLOY
CHSSC, Université de Picardie Jules Verne, Amiens, France

2.1. Introduction

Tracing the history of the clinic in the modern age means understanding the change in perspective on the concept of disease. From humoral theory to organic lesions, disease is no longer an imbalance of humors that must be restored to equilibrium but a species in its own right. Once the body became an object of study, disease acquired a morphological entity and doctors sought to explore its inner workings and understand its manifestations and evolution. On the basis of the botanical model, different pathologies were classified according to the organs affected. However, this classification was only possible if doctors paid close attention to the evolution of each disease entity that the diagnosis depended on. Studying bedridden patients became one of the foundations of medical practice. Symptoms have become the signs from which doctors establish a profile of the disease. This chapter attempts to highlight how the concept of disease gradually shed its mystical, even metaphysical, foundations to acquire a positive basis. From the Paracelsian disruption to the creation of hospital medicine after the Revolution of 1789, this chapter highlights the practical and ideological disruptions that led to the birth of the concept of the clinic.

History of Life Sciences and Medicine 2,
coordinated by Céline CHERICI.

2.2. The birth of a concept

2.2.1. *The evolution of the concept of disease*

While the concept of disease was essentially Galenic and rooted in the imbalance around humors during the Middle Ages, this model gradually lost its supremacy and was repeatedly challenged. The Paracelsian model, for example, linked the microcosm to the macrocosm through structural analogies. Human beings reflect the world. For Paracelsus (1496–1541), diseases were beings with their own structure originating from outside the body. He replaced the four Galenic elements with three principles: sulfur (combustible), salt (solid and permanent) and mercury (fluid and changeable). Different pathologies are the result of alterations in these chemical substances. Each disease is a mineral that corresponds to a specific higher mineral. Skin ulcerations, for example, are thought to be caused by salt, which is conceived as a chemical substance with well-defined physiological and pathological roles (Grmek 1997, p. 162).

Over the following century, the concept of disease took two approaches: first, it became iatrochemical, as put forward by Jean-Baptiste van Helmont (1579–1644), who advocated the existence of an archeus, a spiritual principle governing all physiological activities, with disease being caused by the malfunctioning of this principle. Like Paracelsus, van Helmont's conception of disease was not excessive or deficient but corresponded to a particular ideal entity, a kind of spiritual seed that took hold of the human body and structured it. The pathogenesis of diseases was mostly chemical: it was a local disruption of chemical processes, especially fermentation in organs.

Franciscus de le Boë, known as Sylvius (1614–1672), emphasized that the vital process is a fermentation phenomenon. Diseases were classified into two categories according to their alkaline or acidic predominance. Diseases were therefore considered to be disruptions in organic transformations that occurred due to fermentation anomalies or deviations (Grmek 1997, p. 164).

The second approach was "iatromechanistic". The Copernican and Galilean revolutions, Cartesian mechanistic philosophy and the instrumentalization of nature led to a new chapter in the history of philosophy. All these developments led medicine in new directions, with a mechanical explanation of physiological processes and pathological processes. Giovanni Borelli (1609–1679) proposed a general anatomo-mechanical explanation for pestilence and fever. Malignant fevers were believed to originate from toxic exhalations from the earth, which penetrated the body and directly affected the organs. Fever was considered to be a beneficial reaction, consisting of an acceleration of the body's internal movements (Grmek 1997, p. 165).

All diseases can be reduced to alterations in vital movements, particularly abnormalities in the passage of particles through organs. This model of pathology originates from the fact that life is a constant movement of particles and that the body is composed of nerve channels and pores through which these particles circulate.

The modern era saw the emergence of neo-Hippocratism. According to many medical historians (Grmek 1990), this return to Hippocratic views was the result of the rejection of Galenism in university teachings. However, another reason can be put forward: the practices of the father of medicine. In Hippocratic medicine, it is not knowledge acquired from books and scholarship that prevails but rather the bedside experience of sick patients. Hippocratic medicine provides a different view of the human being – no longer solely in relation to the divine but rather to the universe. Humors respond to natural laws: phlegm (cold and damp) is predominant in winter, blood (hot and damp) is predominant in spring and yellow bile (hot and dry) and black bile (cold and dry) are predominant in autumn. Human diseases depend not only on living conditions but also on local natural factors: the direction the city is facing in relation to the winds, the quality of the water and the nature of the soil. These factors have an effect on individuals, who react according to their nature, sex and age. The physician must take these factors into account in his diagnosis and prognosis of the disease. Human beings are in good health when influences are measured and balanced (Hippocrates 1800).

2.2.2. *The physician–patient meeting*

From the 16th century to the 18th century, encounters with patients could take place in two ways: either de visu or by letter. In the first case, the doctor would write a *consilium*, a kind of logbook for the patient, and in the second case, it was more of a memoir written by a general physician, by the patient themselves, or by the patient's relatives. The written form was mainly used by patients who lived far from large cities, where physicians were located. It was also used by physicians who had exhausted all possible remedies for a condition and sought advice from more experienced physicians. This type of consultation was only possible for medical consultations.

Consilia, like memoirs, follow a specific structure. They generally consist of three parts. First, the physician records the consultation and examines the situation.

Next, he describes the patient's name, sex, age, social status and place of birth. He lists all of the patients' symptoms and attempts to define the illness and its causes. Finally, he prescribes remedies, along with various recipes and doses of the necessary ingredients. He also prescribes technical interventions (bloodletting, cauterization and fumigation), indicating order and frequency (Agrimi and Crisciani 1994, p. 27).

When making a diagnosis, the physician draws a list of symptoms and links them to a number of possible causes. This enables the physician to structure the pathological situation by isolating the main disease and any secondary or related conditions (ibid., p. 28). Through *consilia*, physicians are able to break free from the constraints of authoritative theory and act. By being at the patient's bedside, they are able to combine practice and theory. The starting point is no longer the text but the patients themselves. A medicine based on discourse emerges from a medicine based on dialog: a dialog with the body. In his *Consilia*, Jean Fernel, for example, highlights a new step in the art of diagnosis that contributes to the empowerment of the clinic. Disease speaks the language of the body, and it is up to the physician to understand its rationality (Pittion 2002). "The formation of the clinic is linked to the emergence of the medical gaze in the field of signs and symptoms" (Foucault 1963, p. 90). Symptoms cease to be passive natural phenomena and become signs of illness.

2.2.3. *Jean Fernel and the concept of disease*

In 1554, *La Medicina* by physician Jean Fernel (1497–1588) was published and was divided into three parts: *Physiologia*, *Pathologica* and *Thérapeutica*. Fernel's method in *Pathologica* is the same as that in his *Physiologia*, "that is to say, proceeding in order, from the general to the particular, from the whole to the parts" (Fernel 1655, p. 26). With Fernel, we see a new definition of disease, which is particularly related to the body. It "is a certain effect produced on the body by something that alters its disposition against the order of nature" (Fernel 1660, p. 3), and he goes on to assert that "every constitution and disposition of the body is natural or unnatural" (ibid., p. 5). He classifies diseases into two categories: first, those that originate inside the body, such as "gravel, stones or thick, hardened phlegm blocking the neck of the bladder and preventing urine from passing" (ibid., p. 7) – in other words, the organ no longer performs its function; second, diseases that originate outside the body that do not prevent the body from functioning. Fernel's division is essentially clinical and partially etiological for generalized diseases without a specific location (such as fevers) and anatomical for localized diseases.

Thus, distinguishing between the causes of the disease and the disease itself, which impairs organ function, is important. Certain states of lethargy result from various lung conditions. However, the initial symptoms, such as coughing and difficulty breathing, can have various causes. The patient's body is examined in detail, taking into account how the patient feels. Symptoms refer to what can be perceived by the senses, but not all bodily changes necessarily indicate a pathological condition. Fernel is well aware that a symptom can be shared by several conditions. It is therefore important to carefully identify which symptom derives from which disease. With Fernel, the concept of disease is redefined. It is no longer magical, it has shed its religious influence and it is no longer considered divine punishment. Although Fernel remained attached to classical humorism, he differed from Galen in his clear distinction between signs and symptoms and in his very conception of disease: they were not all caused by a mixture of humors, but some were due to matter and others to form.

At the end of the 17th century, several doctors were well aware that to advance the art of medicine, it was necessary to change attitudes and the model on which medical knowledge was based. This was one of the tasks that the Italian Giorgio Baglivi (1668–1707) set himself.

2.2.4. *Gorgio Baglivi and the reshaping of medical art*

In his *De l'accroissement de la médecine pratique* (The growth of practical medicine), Baglivi raises objections against abstract medical theories, which he describes as "beautiful reasoning". In his view, physicians must make accurate observations and, through reasoning, organize them meticulously to formulate aphorisms through successive generalizations that highlight the sequences and correlations observed. He emphasized that observation and the use of reason reveal "[...] the consequences of things to the physician, disclose the principle and cause of diseases to his eyes, and enable him to foresee their course and outcome" (Baglivi 1851, p. 2). The author's main objective is to free medicine from the doctrines and prejudices that have hindered its progress, and his aim is to establish it in an experimental philosophy. As a worthy successor to Francis Bacon, whom he quotes extensively in his *Praxis*, Baglivi emphasizes that the basis of medicine is the observation of nature:

> It is not in the penetration of the human mind that we must seek the cradle of this sublime art: it is in careful, delicate, incessant observation, a kind of spying of nature that creates medicine. (ibid., p. 4)

Similarly, the author attempted to identify the obstacles encountered by medicine before it became a practical science. He identifies six obstacles: contempt for the elders, particularly Hippocrates; constant reliance on medical idols; poor reading of treatises; misinterpretation of authors; the way systems were constructed; and lastly, the neglect of the aphoristic method in the description of nature. In other words, there needs to be a return to what he considers to be true medicine, that of Hippocrates, the physician who brought this discipline to perfection through the history of diseases and therapeutic principles.

According to Baglivi, medicine has two levels: reasoning and observation. The progression of diseases has remained the same for centuries: "The nature of diseases, their progression, and their frequency have not changed" (ibid., p. 87). Rejecting a view of medicine that is full of uncertainty and without foundation, he revamps Hippocrates teachings by highlighting his genius in identifying two types of symptoms in diseases: those that are constant and specific and those that are irregular and common to several pathologies. The former are the characteristic signs of the disease. Some of them are obvious and palpable, whereas others are obscure and difficult to interpret. For the latter, Baglivi believes that they should not be ignored by physicians but instead be studied with natural and vigorous simplicity. The focus is shifting toward a more natural philosophy. It is by studying the different links that nature has with its parts that we can deduce the different relationships between them. It is by experimenting with nature that we come to understand it. Through many observations, physicians can build a solid structure. In describing diseases, he must proceed step by step. First, it is necessary to seek out and describe an infinite number of specific facts related to it. Once the observations have been gathered, it is necessary to deduce indisputable axioms from them and form a clear and precise idea of the theory of the disease in question.

> Observations are like letters of the alphabet; each one has no value or usefulness in itself, but when placed next to each other, united, arranged in a thousand different ways, they eventually become true language, the true expression of nature. (ibid., p. 328)

Observation and reason are inseparable because "reason can never be at odds with experience" (ibid., p. 6). According to the author, reason is paramount; it reveals the consequences of things to the physician, discovers the principles underlying the causes of disease and leads to knowledge. In disease, two things must be distinguished: the history of diseases, which the author calls "primary medicine", and their treatment, "secondary medicine". The former mainly aims to remove all medicine-related theories that are contrary to observation. The latter is based on the first and can be clarified by all other sciences: chemistry, botany, mathematics, etc.

The framework of the medical edifice can only be method: the one proposed by Baglivi is induction.

> To perfect all branches of nature, and especially medicine, there is no better method of reasoning than induction: first, because it follows nature step by step, so to speak, and merges with it; and second, because through this method of reasoning, the smallest discrepancies of the mind become a thousand times clearer and more evident than through any other. (ibid., p. 37)

Given the difficulty of developing clinical medicine that could exploit the knowledge acquired through new anatomy and physiology, Baglivi attempted to create a clinical practice that would encompass listening to and observing the patient while analyzing the disease in the depths of the body (Toscano 2016, p. 7).

2.3. Classification thinking

2.3.1. *Sydenham's clinical concept*

> Aware of the futility of efforts to deduce the nature of diseases from general principles and reduce the essence of diseases to chemical, physical, or spiritual processes, Sydenham started from the clear observation of a practitioner: diseases present themselves in the form of clustered symptoms. (Grmek 1997, p. 167)

Breaking from all existing medical models, the English physician Thomas Sydenham (1624–1689), or the English Hippocrates, as some physicians like to call him, proposed a new concept of disease:

> Disease is nothing more than a force of nature, which, in order to preserve the sick person, does everything it can to expel the morbid matter. (Sydenham 1816, p. 2)

He considers it a beneficial force of nature, which works with all its might to overcome the obstacles that hinder the activity of functions. In the advancement of medicine, two assumptions are necessary: first, to have a history or description of all diseases, which are as accurate and faithful as possible; second, to have a sure and certain method. In the history of diseases, Sydenham presented four rules. The first is to reduce all diseases to precise and determined species, as botanists do when they classify plants. Quite often, the imagination and speculation of authors have impeded medical progress and caused them to confuse different classes of diseases. Some diseases may be of the same type and have the same name, with similar symptoms, yet may prove to be of a very different nature. The second rule is that all philosophical hypotheses must be abandoned and that the smallest details that appear to be clear and natural must be noted. In the third rule, the specific and essential symptoms must be presented separately from the accidental or extraneous symptoms. Accidental symptoms include those that depend not only on the age and temperament of the patient but also on the way the disease is treated. There are certain symptoms that are more an effect of the remedy than the disease itself. The fourth and final rule is that it is necessary to consider the seasons that favor each type of disease. This is therefore a matter of linking the succession of seasons and epidemics. Knowledge of the seasons that produce diseases is very useful to physicians not only for distinguishing the type of disease but also for curing it.

In Sydenham's theory, diseases behave like human beings and appear with a certain frequency and in a specific location. At the same time, he was convinced that the process of pathogenesis was completely incomprehensible and inexplicable. Between 1661 and 1664, he conducted a systematic study of various intermittent and continuous fevers, carefully noting the climatic conditions and all the symptoms inherent to each type of fever. Although Sydenham did not attempt to classify diseases into a comprehensive system, he nevertheless paved the way for natural nosology by considering them constant species.

2.3.2. *Boissier de Sauvages and nosology*

Like most doctors in the 18th century, Boissier de Sauvages (1706–1767) was opposed to medical systems. His criticism was based on the missing link between practice and theory. For him, this explained why medicine had made so little progress. Using fever as an example, he showed that pathologies could be assessed differently within the same school: "One attributes it to irritation and heart palpitations; another to the pressure and distension that this viscera experiences from blood; a third to obstruction of the extremities of the blood vessels [...], etc." (Boissier de Sauvages de Lacroix 1772, p. 66). The French doctor claimed that there was an inconsistency and that even if physicians had different views on fever, they nevertheless used the same remedies.

Inspired by Sydenham and the botanical classification system developed by the Swedish physician Carl Linnaeus, he proposed a method for classifying diseases. He published the first drafts of his theory in 1731 in the *Nouvelles classes de maladies*, which, similar to the botanical classification system, included the genera and species of all diseases, along with their symptoms and indications. He made the following observation:

> We lack a history of diseases, based on nature and without myths. [...] We lack a general history of diseases that mark the nature, growth, and demise of each species. (Boissier de Sauvages de Lacroix 1731, p. iij)

The results of his work were published in 1763 in five volumes entitled *Nosologia methodica sistens morborum classes, genera et species, juxta Sydenhami mentem et Botanicorum ordinem*. His work earned him great renown in Europe and was translated into several languages. He classified diseases by placing them into a hierarchy, ranging from the most general symptoms to the most specific symptoms, comprising a total of 10 classes, 295 genera and 2,400 species. He described three ways of acquiring new knowledge, notably history, philosophy and mathematics:

> History is the knowledge of facts. For example, pleurisy is accompanied by fever, difficulty breathing, coughing, and chest pain. Philosophy is knowledge of causes and principles; thus, someone who knows the causes and principles of the four symptoms that accompany pleurisy has philosophical knowledge of pleurisy, knowing that they result from inflammation of the pleura or lungs. Mathematical knowledge consists in knowing quantities and knowing how to measure them, for example, determining the strength and speed of the pulse, the degree of heat, the intensity of pain, the violence of coughing, and other such symptoms. (ibid., pp. 91–92)

The nosological method proposed by Boissier de Sauvages distinguishes between two types of classification: synoptic and systematic. Synoptic classification divides diseases into two opposing categories: external and internal. Moving away from alphabetical and anatomical classifications, which classify diseases according to their location in the affected organs, and temporary classifications, which classify pathologies according to their frequency, Boissier de Sauvages prefers the systematic method, which groups together diseases with similar and dissimilar appearances. He limits specific diseases to their species and these species to their genus, the genera to orders, and, finally, the orders to classes. The genus provides signs for recognizing the defined thing, for distinguishing it from others, with these signs constituting its definition. Signs are defined as the intrinsic qualities of things. They serve to make them known and to differentiate them.

For clear nosology, certain terms that are common to several subjects or fields, such as physiology, botany, and zoology, should be avoided. Words such as "appetites", "fury", and "passion" should also therefore be avoided. The same applies to barbaric words and compound words. Simple words should be used in the definition of genres. For example, hepatitis is preferred to inflammation of the liver; enteritis is preferred to inflammation of the intestines. Generic names of diseases can be used to designate the species when an epithet is added. This is reminiscent of the binomial classification system established by Carl Linnaeus. In 1731, Boissier de Sauvages stated that the definition of diseases, classes, genera and species must be clear and concise. Disease is defined as "a combination of notable symptoms related to one another" (ibid., p. 321).

The most obvious and consistent symptoms appear first and constitute the essential characteristics of diseases. They can also be called "the characteristics of the disease". There are three kinds of characteristic signs: functions, excretions and qualities. The characteristics of diseases must be determined on the basis of symptoms. A doctor who is thoroughly familiar with the signs of the disorder and the combination of different symptoms will save time and be able to easily navigate the different classes, species and orders of disease.

2.3.3. *Philippe Pinel and the revival of nosography*

Sauvages' nosography and Sydenham's teachings greatly influenced thought on classification in the 18th century. In 1798, Philippe Pinel (1745–1826) published *Nosographie philosophique*, in which he attempted to reconstruct a nosographic system that he had previously considered inadequate. He emphasized that nosography undoubtedly provided a tool for describing, naming and classifying diseases. However, he recommended going further into the study of signs so as not to get lost in nuances and subjectivity. Thus, the nosological distribution and medical nomenclature, particularly in the case of acute diseases, need to be modified. In the case of fevers, for example, he identified several types, such as bilious inflammatory fevers, bilious putrid fevers and *mesenteric* bloody fevers. According to Pinel, one of the mistakes made by nosologists was to classify these fevers according to their complications, which were medically non-sensical. Nosological tables, which are divided into classes, orders, species and genera, resemble "arbitrary and inconsistent classifications, symptomatic conditions mistaken for primary diseases, excessive proliferation of both" (Pinel 1790, introduction, p. V). Pinel highlights six primitive feverish orders: vascular blood disorders; particular irritation of the stomach, duodenum or neighboring parts; irritation of the membranes of the food duct; consideration of changes in these ducts, involving a feeling of weakness or sluggishness around the irritable muscle; a deep irritable and sensitive lesion; and the characteristic traits of the particular circumstances of mortality, contagion and simultaneous glandular involvement (ibid., Introduction, p. XV).

It is important for nosography to accurately describe the diseases studied in relation to their similarities. They form a natural sequence of ideas classified by external signs. With this new method, it would be easier for physicians to classify new pathologies. Pinel's intention was to make medicine scientific and to provide students and young practitioners with a manual that did not yet exist.

As Michel Foucault noted, the clinic appeared as a new profile for the experience of the physician, which could be seen and expressed. A reorganization of bodily space occurred, as did the elements constituting the pathological phenomenon, the definition of a linear series of morbid events and an articulation of disease with the organism. The emergence of the clinic as a historical fact must be identified within the systems of these reorganizations (Foucault 1963). The various symptoms and signs have been translated into a language that the physician can decipher. However, to do so, the physician must learn to listen because the medical gaze is a silent gaze that does not intervene.

2.4. Searching for an effective treatment

The modern era began with the disappearance of certain scourges, such as the plague, but the Old World saw new ones, such as malaria and smallpox, emerge. The development of human and commercial exchanges, wars and travel facilitated the spread of these infections (Mollaret 1997, p. 253). As a result, faced with these new infections, doctors found themselves helpless and had to develop new therapeutic strategies. From the 16th century onward, health and disease were no longer solely a matter of individual experience but also became collective issues. The individual relationship between the patient and the physician, which was aimed at healing, was complemented by a more comprehensive approach focused on prevention (Faure 1997, p. 279).

According to the definition in the *Encyclopédie*, the term "therapeutic" is defined as the art of curing diseases, discovering the most effective curative measures and applying them. Therapeutics is divided into diet, surgery and pharmacy. The term "diet" is to be understood in the sense of a healthy lifestyle (diet, exercise, sleep, etc.). Although this definition from 1751 brought about a rethinking of medical science, at the dawn of the modern era, two approaches to therapeutics remained: the medical approach and the surgical approach. These two approaches adopted different therapeutic strategies. First, the humoral approach focused on the circulation of fluids that constantly move through the body. Excess fluids were expelled during illness by administering healing forces aimed at restoring the lost balance. The use of bloodletting, among other things, helped to rebalance this circulation. Blood was extracted using cupping, leeches or phlebotomy.

Second, the solidist approach was used. This approach aimed to locate and discover lesions by seeking to understand their anatomical structure. The surgical approach remained essentially anatomical, topographical and functional. This separation is said to have resulted from a centuries-old prejudice dating back to the time when medicine was practiced by religious figures, but it is also said to have been social in nature owing to the nobility's contempt for manual labor (Henry Valési, 1938). This implies that the physician was a scholar and that the surgeon was a craftsman. In some faculties, such as that of Paris, students were expressly advised not to use their hands under the penalty of fines.

In the 17th and 18th centuries, aided by anatomical pathology and experimental physiopathology, in other words, the experimental study of disease processes, surgery evolved from an empirical practice to a technique based on precise knowledge. This enabled doctors to understand the dynamic nature of lesions. Surgery thus became a clinical activity in its own right. Surgeons play an increasingly significant role within the medical profession because of their training and knowledge of anatomy and pathology. During the Enlightenment, many advances were made in the field of arterial ligation, particularly owing to the work of surgeons such as Jean-Louis Petit (1674–1750) and Pierre Dionis (1643–1718). The year 1731 saw the creation of the Academy of Surgery in Paris.

At the beginning of the modern era, diseases were treated by people with a wide range of professional training and social status, allowing a certain amount of charlatanism to develop. Many skilled charlatans turned healing into a public spectacle. While urban areas benefited from numerous doctors and surgeons trained at universities, rural areas had their own recognized healers.

To better understand the therapeutic strategies of the 16th and 17th centuries, it is important to consider the wide variety of people working in the field of therapy and to highlight the disorganization of the medical system, which was inaccessible and expensive. For example, consultations were sometimes billed between 6 and 24 pounds, which represented 18, 36 or 72 days' worth of wages for low-skilled workers (Costes 2014, p. 52). This situation encouraged a certain type of home medicine without the need for doctors. The Age of Enlightenment saw a pocket medical dictionary boom, full of practical advice on how to stay healthy. For example, some dictionaries included a table of weights and measures with conversions: one pint is 2 pounds, one pound is 16 ounces, one ounce is 8 gros and one gros or dram is three scruples or 72 grains of barley (Rey 1991, p. 13). In *L'avis du peuple sur sa santé*, published in 1761, the Swiss Samuel Tissot offered rudimentary advice to peasants on hygiene and self-medication, fearing that the lack of care in rural areas would lead to their desertification.

2.5. The anatomo-clinical method

> Until the 16th century, doctors had few opportunities to broaden their knowledge and clinical experience. Private practice was limited to the upper classes of society, and hospital consultations were sporadic. Autopsies were rare and limited to cases involving forensic medicine or public health. (Risse 1997, p. 177)

At the beginning of the 16th century, there were no clinical works as we understand them today. Clinical descriptions were included in medical treatises. The Florentine physician Antonio Benivieni (1443–1502) was one of the first to attempt to correlate clinical observations with autopsy findings. His objective was to find the hidden causes of disease, and he gained insight into complex pathological processes, such as the transmission of syphilis from mother to fetus and inflammation of the peritoneum following intestinal perforation. However, he had too few observations to form a system. Even though he tried to bring something new to the table, Benivieni remained very attached to medieval humorism.

In the 17th century, anatomical research was propelled by the 1679 publication of *Sepulchetum sive Anatomica practica* by Swiss physician Théophile Bonet (1620–1689), which was revised and republished in 1700, 11 years after his death. *Sepulchretum* is considered the first work on anatomical pathology, although it contains a large number of observations borrowed from other authors. In his *Observations et historiques chirurgiques* (Bonet 1670), Bonet republished or translated the observations of other practitioners, such as Thomas Bartholin, Guillaume Baillou, and Antonio Benivieni, and included his own observations. The author advised physicians to compare the cases they encountered in their practice with those described in his book to improve their diagnoses. The author was convinced that dissections could help discover the causes of diseases. He thus listed nearly 2,934 observations from 470 authors, with diseases classified in order. In 1682, he published *Mercurius compilatus*, a medical encyclopedia that described and classified diseases in alphabetical order.

In 1761, Giovanni Baptista Morgagni (1682–1771) published *De sedibus et causis moborum par anatomen indignatis*. Although *De sedibus* was initially intended to correct the errors of *Sepulchretum*, it became a collection of clinical medicine. Despite recognizing the significance of Bonet's work, he remained critical, noting its lack of rigor, particularly in the organization of the various tables. Considering these tables to be poorly arranged, he showed that they referred to several entities, which made it necessary to constantly navigate through the work. Nevertheless, he follows Bonet's example in their constitution, beginning with diseases affecting the head. He also noted incomplete observations, which may cause doubt in the minds of physicians. Morgagni also partly repeats the observations of his master Valsalva, as well as those of some of his friends, and adds his own observations. This treatise is a major work, illustrated by observations gathered throughout his life.

During the 18th century, many treatises on anatomical pathology appeared, such as that by the Frenchman Joseph Lieutaud (1703–1780), entitled *Historica anatomica medica* (Lieutaud 1767), modeled on *Sepulchretum*, in which he summarized 3,500 clinical cases accompanied by autopsy reports. What was previously invisible became visible, insofar as a causal link was directly established between symptoms and organ damage. In 1710, François Pourfour du Petit (1664–1741) investigated the links between a head injury and paralysis on the opposite side of the injury. Through anatomopathological research and vivisection experiments, he was able to describe the alteration of striated bodies and paralysis. Anatomopathological research was part of the process of gathering evidence. While many physicians were opposed to speculative medicine, this causal link introduced a degree of rationality into medicine. Practitioners were no longer solely relying on the observation of bedridden patients, as Baglivi had emphasized, but were taking direct action on living beings by subjecting them to experimental protocols. The Age of Enlightenment ended with the discovery of tissue pathologies, the main advocate of which was Xavier Bichat (1771–1802), whose aim was to establish a defined relationship between tissue alterations and functional alterations. According to him, tissue lesions determine the appearance of clinical symptoms.

2.6. Institutionalization of the clinic: the hospital

During the 18th century, there was a shift from species medicine, which was the natural setting for disease, in other words, family, to class medicine, which was institutionalized medicine:

> Clinical observation requires the organization of two interrelated areas: the hospital domain and the educational domain. (Foucault 1963, p. 109)

Foucault also mentioned that the hospital is the place where pathology naturally appears. Notably, under the Old Regime, hospitals were sometimes used to cure sick people or train doctors but were rarely used to advance knowledge. Hospitals were crowded with children, the elderly and vagrants, which could lead to the death of even the most resilient individuals. In addition, there were too few beds to guarantee clinical training, an issue that concerned theorists of new medicine after the revolution. Thus, how many beds were needed to guarantee quality teaching? Opinions were divided: while some clinicians believed that few beds were needed, others believed that a certain number of beds were necessary to treat a range of conditions. It was accepted that 20–30 beds was the ideal number. Under the Old Regime, there were initiatives that attempted to adopt an educational approach, such as that of Pierre Hunauld (1664–1728) in 1716. Under his direction as dean of the faculty of medicine in Angers, he managed to gather a few beds in the city's hospitals so that the faculty's professors could teach there.

However, these initiatives remained rare in France, where medicine was still seeking to codify itself. Physicians were gradually realizing that the texts of their predecessors were no longer sufficient to explain all the diseases. The new medicine that was emerging took two forms: university education and observation of bedridden patients.

Hermann Boerhaave (1668–1738) is credited with the initiative of teaching from the patient's bedside. As head of clinical medicine in Leiden in 1715, he had a dozen beds installed with selected patients, both men and women. These were referred to as the men's clinic and the women's clinic. The aim of the lessons was to assess the students' knowledge and familiarize them with their future profession under the guidance of a supervising physician. Boerhaave's clinic was strongly influenced by the work of Sydenham. He even proposed a classification of diseases on the basis of anatomical and clinical criteria, as well as a hierarchy of clinical signs in disease, which would form the basis of diagnosis. This method was emulated: Boerhaave's students introduced this practice in many countries. However, it must be acknowledged that in France, under the Old Regime, the organization of clinical teaching remained very limited. Although the Edict of Marly in 1707 aimed to reorganize medical education and recommended the creation of a head of practical medicine to teach students how to practice and consult, it was not until the Revolution that clinical education was organized.

In Edinburgh, from 1748 onward, Boerhaave's model was adopted by the doctors Daniel Rutherford (1749–1819) and William Cullen (1710–1790). The Scottish clinical model was recognized throughout Europe for its effectiveness. In 1751, it was Austria's turn to reform its medical education under the incentive of Gerard van Swieten (1700–1772), one of Boerhaave's students. With the emergence of clinical schools, medical education reform took place. In Pavia, Samuel Tissot (1728–1798) also reorganized medical education in 1781. Unlike Boerhaave, Tissot taught in French. He believed that it was important for education to be taught in the vernacular language. In his *Essai sur les moyens de perfectionner les études de médecin* (Essay on the ways of perfecting medical studies) (Tissot 1785), the author outlined what he believed to be the ideal clinical teaching method.

It was in postrevolutionary France that real clinical teaching was organized. On 14 Frimaire Year III (December 4, 1794), under the incentive of Antoine François Fourcroy (1755–1809), who was then tasked by the convention with presenting a report on the creation of health schools in Paris, the first measures were implemented. The aim of his various reports was to rebuild medical practice. The decree of August 18, 1792, had abolished universities, faculties and learned societies, and one of the consequences was that health schools began to accept physicians and surgeons. Another consequence was the establishment of medical anarchy, where "the most dangerous empiricism and the most shameless charlatanism abused gullibility and good faith everywhere" (Fourcroy 1814, p. 446). To combat this slack, on 16 Frimaire Year III (December 6, 1794), 4 years after the abolition of the faculties, three health schools were created: those of Paris, Montpellier and Strasbourg.

As early as 1790, in his *Nouveau plan de la constitution en médecine* (New plan for the constitution of medicine), Félix Vicq d'Azyr (1748–1794) argued for a closer relationship between medicine and surgery because "it is by restoring surgery to medicine and medicine to surgery that we come closer to nature" (Vicq d'Azyr 1790, p. 6). In this new teaching, students had to master both internal (medical) and external (surgical) clinical practices. Teaching also had to cover knowledge of animal economics, namely, the study of the anatomy, chemistry and physiology or physics of healthy humans; simple and complex substances, whose action influences said economy; the choice of means to keep the body healthy; the examination of various diseases; their symptoms; their indications; and the remedies to combat them and the history of science itself. The author of the new plan continues by stating that "physics, natural history, anatomy, chemistry, and clinical observation are the foundations on which the entire edifice of the medical institution must be built" (ibid., p. 21). He emphasized that in the organization of health schools, practical training should be combined with theoretical precepts, and in his view, bedside observation should be one of the most important parts of this rigorous education in prerevolutionary medicine. To this end, it was necessary to recruit knowledgeable professors who would be supported by assistants. This reform was characterized by a rebalancing of medicine around clinical practice. Four years later, Fourcroy took up his friend's observations and conclusions in his own report.

Once practical experience gained from the patient himself has been defined, there is a necessity to link specific knowledge to a general system of knowledge (Foucault 1963, p. 71). The hospital becomes educational when the sick are separated from the poor (Keel 1985, p. 186). The hospital situation at the end of the Old Regime and in the early years of the New Regime underwent major changes following the introduction of five principles: the improvement and enlargement of the old hospitals; the establishment of a clear distinction between medical institutions (hospitals) and philanthropic institutions (for children, the elderly, etc.) and prisons; the construction of new, smaller hospitals; the seizing of convent hospitals by the state; and the state's takeover of hospitals and the centralization of their administration (Ackerknecht 1986, p. 32). It should be remembered at the same time that hospital medicine developed outside the university system. In the second half of the 18th century, hospitals became the new venue for medical practice and research before clinical teaching was incorporated into the university system (Keel 1985, p. 188) Advances in clinical teaching were largely due to developments in surgery. Pierre Joseph Desault (1738–1795), for example, a surgeon by profession, was appointed a professor of clinical medicine at Hôtel-Dieu in 1787. The postrevolutionary clinic also saw the specialization of certain hospitals: Saint-Louis Hospital, a former quarantine station for plague victims, built in 1604 during the reign of Henry IV, specialized in skin diseases; Midi Hospital became a hospital for venereal diseases; Maternité, a former convent in Port-Royal, was converted into a maternity hospital in 1795; and Salpêtrière, which opened in 1656 as a former prison, was converted into an institution for sick, incurable or insane women in 1795 (Ackerknecht 1986, p. 37).

2.7. Conclusion

Tracing the history of the clinic in the modern era means highlighting the different stages of medicine from the Renaissance to the Enlightenment. The term "clinic" was forged through a series of failures and successes. It became more precise with each ideological shift and advance in medical examination. The progress of the clinic lies in overcoming its failures, owing to which it was able to establish itself as an autonomous activity. From symptom to organic alteration, it is a journey that needs to be understood in its entirety. Thinking about the clinic requires thinking about the evolution of the concept of disease, which is its main objective. In the wake of the Revolution, the combination of surgery and clinical practice brought about broad and profound clinical observations.

2.8. References

Ackerknecht, E. (1986). *La médecine hospitalière à Paris (1794-1848).* Payot, Paris.

Agrimi, J. and Crisciani, C. (1994). *Les consilia médicaux*, C. Viola (trans.). Brepols, Turnhout.

Baglivi, G. (1851). *De l'accroissement de la médecine pratique*, J. Bouchet (trans.). Labé, Paris.

Boissier de Sauvages de Lacroix, F. (1731). *Nouvelles classes de maladies, qui dans un ordre semblable à celui des botanistes, comprennent les genres & les espèces de toutes les maladies, avec leurs signes & leurs indications*. D'Avanville, Avignon.

Boissier de Sauvages de Lacroix, F. (1772). *Nosologie méthodique ou distribution des maladies*, volume 1. Jean-Marie Bruyset, Lyon.

Bonet, T. (1670). *Observations et histoires chirurgiques : tirées des oeuvres latines des plus renommés praticiens de ce temps, par un docteur médecin et comprises en douze centuries avec quatre indices, l'un des autheurs, le second des chapitres ; le troisième des matières & le quatrième des médicaments*. P. Chouët, Geneva.

Coste, J. (2014). *Les écrits de la souffrance : la consultation médicale en France (1550-1825).* Champ Vallon, Ceyzérieu.

Faure, O. (1997). Les stratégies sanitaires. In *Histoire de la pensée médicale en Occident, 2. De la Renaissance aux Lumières*, Grmek, M. (ed.). Le Seuil, Paris.

Fernel, J. (1660). *La pathologie de Jean Fernel, premier médecin de Henry II. roy de France. Ouvrage tres utile à tous ceux qui s'appliquent à la connoissance du corps humain. Mis en françois, par A. D. M. docteur en médecine*, 2nd edition. Mourer, Cadet, Libraire et Imprimeur, Lausanne.

Fernel, J. (1655). *La pathologie ou le discours sur les maladies*, C. de Saint-Germain (trans.). Jean Guignard le jeune, Paris.

Fernel, J. (1665). *Les VII livres de la physiologie*, C. de Saint-Germain (trans.). Jean Guignard le jeune, Paris.

Foucault, M. (1963). *La naissance de la clinique.* PUF, Paris.

Fourcroy, A.F. (1814). *Exposé et motifs des lois sur l'exercice de la médecine.* In *Recueils de lois et de règlements concernant l'instruction publique*, volume II. Brunot-Labbé, Paris.

Grmek, M. (1990). *La première révolution biologique*. Payot, Paris.

Grmek, M. (1997). Le concept de maladie. In *Histoire de la pensée médicale en occident, 2. De la Renaissance aux Lumières*, Grmek, M. (ed.). Le Seuil, Paris.

Hippocrates (1800). *Des airs, des eaux et des lieux*, Coray (trans.). Baudelot et Eberhart, Paris.

Keel, O. (1985). La problématique institutionnelle de la clinique de la fin du XVIIIe siècle aux années de la Restauration I. *Bulletin canadien d'histoire de la médecine*, 2, 183–206.

Lieutaud, J. (1767). *Historia Anatomica-medica sistens numerosissima Cadaverum humanorum Extispicia, quibus in apricum venit genuina morborum sedes ; horumque reserantur causae, vel patent effectus ; auctore Josepho Lieutaud ; recensuit et suas observationes adjecit A*. Portal, Vincent, Paris.

Mallaret, H. (1997). Les grands fléaux. In *Histoire de la pensée médicale en occident, 2. De la Renaissance aux Lumières*, Grmek, M. (ed.). Le Seuil, Paris.

Pinel, P. (1790). *La nosographie philosophique*. Maradan, Paris.

Pittion, J.P. (2002). Médecine et médecins au 16e siècle. Jean Fernel (1497-1558), médecin d'Henri II : vie et œuvre. In *Actes du IXe colloque du Puy-en-Velay*, Viallon-Schoneveld, M. (ed.). Presses Universitaires de Saint-Etienne, Saint-Etienne.

Rey, R. (1991). La vulgarisation médicale au XVIIIe siècle : le cas des dictionnaires portatifs de santé. *Revue d'histoire des sciences*, 44(3/4), 413–433.

Risse, G. (1997). La synthèse entre l'anatomie et la clinique. In *Histoire de la pensée médicale en occident, 2. De la Renaissance aux Lumières*, Grmek, M. (ed.). Le Seuil, Paris.

Sydenham, T. (1816). *Médecine pratique*, A.F. Jault (trans.). Sevalle, Montpellier.

Tissot, S. (1785). *Essai sur les moyens de perfectionner les études de médecine*, Mourer, Cadet, Libraire et Imprimeur, Lausanne.

Toscano, A. (2016). Gorgio Baglivi, un médecin en Italie, Arménie et Croatie. *Scienze e Ricerche*, 90, 10–17.

Vicq d'Azyr, F. (1790). *Nouveau plan de la constitution de la médecine*. Société Royale de Médecine, Paris.

3

Experimentation on Living Beings in the Modern Era

Jean-Claude DUPONT
CHSSC, Université de Picardie Jules Verne, Amiens, France

3.1. Introduction

The exploration of life in the modern era follows many paths. It involves supplementing renewed knowledge of living forms (anatomy) with new details, exploring the intimate texture of what can be observed with the naked eye (micrography) and classifying "newly discovered" living forms using microscopic instruments (taxonomy). However, it also involves analyzing the functions that correspond to each of the anatomical forms described (physiology), understanding their genesis (embryology) and the material transformations they undergo (chemistry). Most of these avenues of exploration are thus likely to receive instrumental support or even prove inseparable from experimentation itself. In the fields of physiology, embryology and chemistry, experimentation on living organisms thus demonstrates scientific productivity, including a particular richness of philosophical and epistemological aspects. In this chapter, we propose to follow the thread of the major factual and conceptual advances in animal functions that have resulted from experiments and the controversies they have generated[1].

1 In addition to the primary sources referenced in the text, a selection of secondary sources is provided at the end of the chapter for further reading.

History of Life Sciences and Medicine 2,
coordinated by Céline CHERICI.

3.2. The emblematic case of blood circulation

3.2.1. *Blood movements before Harvey*

Through his discovery of blood circulation and the method he used, William Harvey was a key figure in modern physiological experimentation, ultimately allowing us to explore its characteristics.

However, the experiment-based study of blood movement is not recent. Galen (1538), who had an authoritative doctrine, described it in his *De usu partium corporis humani*, which was notably republished by Sylvius. The liver produces blood from food that is purified by the stomach and transferred by the portal vein. Upon leaving the liver, part of this blood reaches the heart via the vena cava, and the rest travels to the lower parts of the body. The heart has two chambers, the right ventricle (RV) and the left ventricle (LV), with the atria more or less merging with the veins. Once in the heart, some of the blood supplies the head, whereas another part enters the RV. Some of the ventricular blood supplies the lungs, whereas the rest passes from the RV to the LV through openings that Galen claimed to have observed in the interventricular wall. The blood produced by the liver and distributed throughout the venous system therefore enters the arterial system through these interventricular pores but also indirectly through anastomoses between veins and arteries located in the lungs. The air taken in by the lungs is brought to the LV. This air and pulmonary blood mix with the blood coming from the RV through the interventricular openings. The physiological heart–artery–lung system thus ensures the production of vital heat (in the heart), its moderation (in the lungs) and its distribution to different parts of the body (through the arteries) carrying vital pneuma, which is produced in the heart from its own heat, blood and air taken from the lungs. The "pneumatized warm blood" leaves the LV through the aorta, which distributes it to different parts of the body through various arteries. In this system, there is communication between the venous and arterial systems, but blood movement is not circular. Between the heart and the lungs, blood moves back and forth through a motion linked to breathing: aspiration of blood during inhalation and reflux of blood during exhalation. Only the left side of the heart has the necessary pumping ability to move blood to peripheral tissues, where it is consumed and permeates the flesh.

Galen justified this system with various anatomical considerations related to the heart and blood vessels: the ventricles, membranous structures (heart valves) and porous structures determine the movement of blood, as does the anatomical position of the blood vessels in relation to the heart and liver. To this was added certain physiological observations, such as the pulsing nature of the heart and vessels, the effect of ligating arteries and the color of the blood.

The remaining history of blood movement corresponds to the gradual dismantling of Galen's physiological doctrine. Andreas Vesalius questioned the existence of interventricular pores in the 1555 edition of his *De humani corpori fabrica* (the first edition dates from 1543). The interventricular pathway was thus blocked. Michael Servetus confirmed the passage of blood through the lungs and its return to the LV, thereby also questioning the existence of interventricular pores while opening the pulmonary pathway. His *Christianismi Restitutio* (Servetus 1553), a theological work that earned him the contempt of John Calvin, developed anatomical arguments, such as the "diverse conjunction and communication" of the "arterial vein" and the "venous artery" within the pulmonary substance, the considerable size of the arterial vein, which would not be justified if the blood were only used to nourish the lungs, the air mixed with blood that passes from the lungs into the heart, indicating that the mixture must have taken place in the lungs, the insufficient capacity of the LV for such a large and abundant discharge, and, last, the interventricular wall, which is devoid of "blood vessels and faculties", rendering it unsuitable for this communication. For all these reasons, it was clear that, just as it passes through the liver from the portal vein to the vena cava, blood also passes through the lungs. This is the first description of small circulation, at least in Europe. Westerners may have been aware of the writings of a 13th-century Arab scholar, Ibn al-Nafis (Damascus), particularly his *Commentary on Anatomy in Avicenna's Canon*, which was translated into Latin by Andrea Alpago in 1547.

Servetus shared a certain way of looking at visible reality with Galen, agreeing with him that there was a teleological relevance between anatomical form and function that was immediately perceptible and understandable. He saw his own doctrine as a necessary addition to Galen's doctrine. However, owing to Realdo Colombo (Venice), pulmonary circulation has become widely known, and it was his work, *De re anatomica*, that William Harvey referred to (Colombo 1559). Colombo wrongly claimed that no one before him had observed or described this blood circulation. The reason for this is not that his student Juan Valverde had previously made the observation (Valverde 1556), since Valverde himself acknowledged that he had come to the same conclusion with Columbus, but simply because Servetus's work was burned after his execution. With virtually no distribution of the work, it was not made public and medically assessed until the end of the 17th century and the beginning of the 18th century, when Servetus's importance was recognized. Cesalpino (1571) is said to have been the first to use the word "circulation", noting that in the veins, blood moves toward the heart, with the vein protruding below the tourniquet and not above. The centripetal movement of blood in the veins was later confirmed by Girolamo Fabrizi d'Acquapendente, Harvey's teacher, who had been studying venous valves since 1574. These valves are membranes arranged at regular intervals inside the veins that allow blood to flow in one direction and prevent it from passing entirely into the lower extremities of the body (d'Acquapendente 1603).

3.2.2. *Harvey's contribution*

The link with Galen was completely severed with William Harvey's famous work, *Exercitatio anatomica de motu cordis et sanguinis in animalibus* (Harvey 1628). Harvey defended the theory of large circulation, according to which blood pumped by the LV into the aorta returns to the right atrium through the vena cavae. Harvey supported his demonstration with a synthesis of data from his predecessors, to which he added his own findings from physiological observation in humans (arms compressed by tourniquets), anatomical dissection and animal experimentation.

Animal vivisection shows that when the beating heart is touched, it hardens as it contracts: the heart is a hollow muscle, similar to the muscles of the skeleton. The contraction of the heart is synchronous with the distension of the arteries. When an artery is pierced, blood spurts intermittently and simultaneously with the contractions of the heart before the animal dies: the heart acts like a hollow muscle pump. The contraction of the heart begins in the atria, then the contraction of the atria pushes the blood into the ventricles, and finally, the contraction of the ventricles pushes the blood into the arteries. Harvey also drew on his knowledge of the anatomy of the membranous structures located in the veins (venous valves), the membranes separating the atria and ventricles (heart valves) and the membranes located at the point where the arteries branch off from the heart (aortic valves). These valves allow blood to flow in one direction and prevent backflow. The Galenic system was thus challenged quantitatively through a threefold demonstration: (1) the amount of blood passing from the vena cava to the heart is much greater than the amount of food ingested; (2) the amount of blood circulating in the tissues is much greater than that needed for their nutrition; and (3) blood returns to the heart through the veins and only through them. Harvey used measurement and weight calculations to estimate the amount of blood passing through the heart at a given time. The intracardiac blood volume was measured as two ounces (approximately 56 g). It empties with each beat (72 beats/min). The amount of blood passing through the heart is $2 \times 72 = 144$ ounces/min, or 8,640 ounces/h (approximately 245 kg), which is three times the weight of a man. Where does all this blood go in such a short time? If it is produced in one hour by the liver from food, it must be absorbed at the same time into the tissues it nourishes. Alternatively, there are porous capillaries at the ends of the arteries connected to the capillaries at the ends of the veins. The heart sends blood through these pores into the tissues, where it passes into the veins, through which it returns to the heart. This is confirmed by the tourniquet experiment: the flow of blood in the veins can be observed as the tourniquet is loosened. The system of superficial veins is the structure through which blood is returned.

3.2.3. *Harvey's innovation and tradition*

Harvey thus overturned the entire field of Galenic physiology by proposing a new way of looking at vital movements, namely, the movements of fluids within living matter. There are also other fundamental differences with Galen. Galen is considered vital for the transformation of substances under the effects of natural faculties, which specialize in a particular transformation. Harvey rejected or delayed the examination of substantial transformations in physiological processes, which he mostly interpreted as simple passive and local displacements. Furthermore, Galen considered that the relationships between anatomical form and function were related expressions of the specific nature of the living being in question, which were simultaneously perceptible without one taking precedence over the other. Harvey saw function as a consequence of anatomical form (structure), which is not easy to achieve without experience.

This reliance on experience is precisely what makes Harvey so modern. In terms of method, Galen and Harvey agreed on the necessity of sensory observation, but Galen, who was more optimistic, believed that it enabled scholars to discover the "how" and "why" of nature, since, as we have previously mentioned, form and function are naturally aligned, with experimentation only being used to verify this. Given the enormous complexity of nature, Harvey felt an absolute need for an appropriate method to understand functions, a "measuring" method that was a middle ground between the systematic empiricism of Francis Bacon and the mathematical and controlled experimentation of René Descartes and Galileo Galilei. Nevertheless, his research program was still based on the medieval triad of *auctoritas*, *ratio* and *experimentum*. In the first paragraph of the preface to his book, he explained that to address a problem, one would first turn to the classics and authorities, reflect on them and then gather personal experiences on the subject. He was therefore not entirely a methodical experimenter in the contemporary sense.

Furthermore, like all Renaissance scholars, Harvey was not monolithic. Aside from his use of measurement and calculation, he remained very close to the Aristotelian and Galenic spirit. When he argued against Gaspare Aselli, the discoverer of the chyliferous vessels (Aselli 1627), stating that it was not necessary to acknowledge the existence of separate vessels for blood and chyle, in other words, the idea of lymphatic circulation, he based his assertions on a chain of ideas rather than on experience. When he compares the circulation of blood in the body to the cycle of water, which, according to Aristotle's *Meteorology*, rises as steam toward the sky through the sun before falling back to earth as rain, he uses the analogy between the microcosm and macrocosm, between the heart and the sun, similar to Aristotle. Similarly, there is no break with the Galenic and Aristotelian tradition when he believes in natural purpose, in the sense of "supreme providence of nature", which "does nothing without reason and takes all needs into account", and which assigns "use and utility" to all parts of the human body. Harvey thus seeks the ultimate purpose of cardiac function and circulation. He accepts the dogma of the heart as the source of body heat. The heart is therefore both a pump and a vital center, producing vital spirit and heat. Respiration is used to cool the heart and blood, just as the latter tends to cool when it comes into contact with the extremities of the body. It is necessary for the blood to return to the heart to restore its heat and spirit, which would otherwise cool down and lose its spirit, such as in dead bodies.

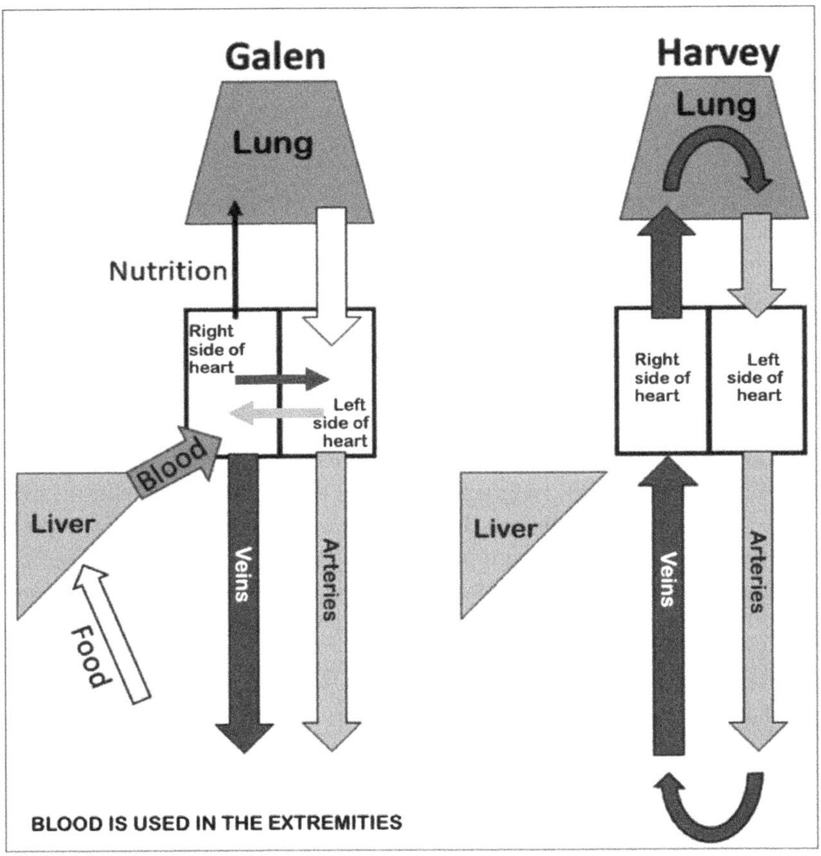

Figure 3.1. *The comparative movements of blood according to Galen and Harvey*

3.2.4. *Controversies over experimental contributions and the usefulness of polemics*

Harvey's doctrine sparked a dispute between his supporters, known as "circulators", and physicians who were loyal to Galenism, known as "anticirculators". Among the numerous latter (including James Primrose, Emilio Parisano and Caspar Hofmann), Harvey agreed to respond only to Jean Riolan (the Younger), with whom he engaged in more or less courteous exchanges. Riolan acknowledged a large amount of circulation but believed in the passage of the interventricular septum through the pores and denied pulmonary circulation. Furthermore, how could the nutrition of organs be explained if the blood flowed so rapidly (Riolan 1649)? Among the "circulators" were Franciscus de le Boë (Sylvius), but above all, René Descartes, who supported Harvey's *Traité de l'homme*, written in the early 1630s, despite some points of disagreement. In line with the Aristotelian tradition, Descartes considered that the very high heat of the heart caused boiling, which was responsible for diastole. The movement of blood was therefore conceived in a thermomechanical way (Descartes 1664). Nicolas Sténon rejected Harvey's analogy between the heart and the sun. The heart was a simple muscle, and the primary agents of movement were bundles of muscle fibers (Sténon 1664).

The dispute on circulation was not merely a battle of posturing between the older generation and modernists; it reflected real concerns. In fact, circulation has raised many new theoretical and experimental questions, and Harvey's work thus sets physiology on a variety of productive paths. Harvey, for example, assumed the existence of porous capillaries connecting arteries and veins, although he did not visualize them. The microscopic anatomy subsequently overturned this situation. The discovery of these arteriovenous structures in the lungs and other parts of the body using the microscope would later reveal the anatomy of the system (Malpighi 1661). If there is arteriovenous communication, how can we explain the differences in color between blood from veins and arteries? Circulation has led to the search for a physiological explanation for respiratory processes, as we shall later discover. The topic of cardiac fiber contraction has led to new research on local movement and muscle function (myology, cardiology). Riolan's objection to the overly rapid hydrodynamics of circulation led to the idea of interstitial plasma. Ultimately, the question of lymphatic circulation led to research into the origin of blood and the role of the liver.

Before Harvey, anatomists from the Renaissance period identified lacteal vessels (chyliferous vessels), which were discovered in dogs by Aselli (1627) and in humans by de Peiresc (1634). In accordance with the findings of Galenic dogma, these authors believed that these vessels emptied into the liver. Pecquet (1651) described the thoracic duct in dogs and demonstrated that it originated in the mesenteric ganglia and ended in the left subclavian vein. The condyle is absorbed by special lacteal vessels, which do not lead to the liver but rather directly to the venous system. This was followed by the discovery of lymphatic vessels by Rudbeck (1653) and Bartholin (1652) and precise descriptions of their valves by Ruysch (1665). Lymphatic circulation complements blood circulation, despite Harvey's reservations. However, above all, the liver was stripped of the considerable role that was given to it by Galen's physiology: it was no longer *fons venarum*, the source of the main veins of the body, and could therefore no longer be considered *fons sanguinis*. The circulation dispute intensified in the mid-17th century through another dispute between the "defenders" and "enemies" of the liver. If it was no longer the source of blood, the question of its usefulness arose, and it was clear that the entire physiology of the liver had to be rebuilt. Nevertheless, the course followed by the two main fluids of the animal, blood and lymph, had been clarified.

3.3. The origins of the experimental exploration of living beings

3.3.1. *The transformation of anatomical practices: from anatomy to physiology*

Harvey's contribution was the most significant achievement in modern physiology, overshadowing other contributions for many years. Harvey was both the cause and consequence of a profound movement, namely, the implementation of Vesalius' *Fabrica* and the construction of this vivid anatomy that would become the emerging discipline of physiology, leading to the establishment of an increasingly autonomous discipline.

Jean Fernel was the first author of a work that included the French word "physiologie" in the title, namely, a translation from Latin of the first part of his *Medicina*, published in 1655 under the title *Les sept livres de la physiologie* (Fernel 1554, 1655). However, Fernel, who describes the different parts of the body and their uses, the various elements and temperaments, the spirits and internal heat, the functions and humors, and generation, remained deeply committed to Galenism, even though he intended to reform it. He was part of medical humanism and ancient and medieval medical tradition, which set him apart from Renaissance scholars such as Vesalius, whose work *Fabrica* was still two-sided: one side was modern and anatomical, describing structures in the Renaissance style, and the other was ancient and physiological, functional, inspired by Galen's legacy.

In the 17th century, under Harvey's influence, the situation changed profoundly. The passion for anatomy remained strong, but the detailed and familiar knowledge of the human body would require it to be understood in a new way, from the perspective of the role of its structures in the performance of specific physiological functions. In the architecture of the parts of the body, the aim was now to understand vital movement: anatomy still provided the dissector with the subject of his investigation, but anatomical discoveries were interpreted from the perspective of the movement that the revealed structure could perform. The general tendency, in the Baroque period, to understand nature dynamically (dynamism), which replaced the Renaissance's static and architectural view of the world, was also expressed in the body, whose representation was transformed. Harvey is symbolic of this new anatomical–physiological spirit. Physiology is initially an animated anatomy, a shift in Vesalius' architectural anatomy. According to Harvey, it is the knowledge of this movement, which is still a spatial displacement (*motus localis*), that is the object of physiology.

From the Baroque period to the Enlightenment, the trend was to make this physiological research an increasingly autonomous activity, placing less emphasis on the anatomical structures themselves than on the decisive role they seemed to play in relation to functions. Anatomical practices changed accordingly. Admittedly, in the 17th century, additional details concerning the skeleton and the muscular system were added to *Fabrica* – a requirement respected by anatomists. However, particular attention was given to parts of the body where local movement was most visible, such as the vessels, through the injection of hardening and colored substances. Frederik Ruysch explored the coronary, bronchial, cerebral and placental vessels. Only the coloring of the lymphatic vessels made it possible to identify them easily and dissect them. For this reason, their injection remained mandatory to differentiate them from other vessels (Ruysch 1665). Anton Nuck developed a device for injecting mercury into vessels. After the lymphatic collector was located, a glass or copper tube was used to blow air into the vessel. Once located and dilated, the vessel was catheterized and injected (Nuck 1691). In addition to the pathways of vessels and the fluids they contain, this period involves the description of dozens of experiments on living beings, the contents of sacs and glands, the liver, excretory channels, the location and role of specific muscles, the nervous system that appears to be at the origin of this movement and the varying stimuli to which organs are sensitive, depending on their nature. In medicine, researchers have investigated the presence or absence of particular elements in health and disease, the natural stages of healing damaged parts, the nature and differences between the parts that make up organs, and the action of drugs and poisons. These were described in works such as those by Pecquet (1651), de Graaf (1668), Dionis (1690), von Haller (1752),

Monro (1785) and Hunter (1794)[2]. This anatomy, with its practical physiological objectives, cruelly involved the vivisection of animals, but some of the experiments could also be carried out on cadavers using observation with or without the aid of a magnifying glass or, possibly, a microscope, which remains a controversial instrument (Ratcliff 2009).

3.3.2. *Explanatory systems*

This tendency toward the autonomization of physiology, relative to anatomy, is expressed through the use of explanatory systems specific to physiology, which organize concrete knowledge and integrate it according to a general view of bodily functioning. The reference here is René Descartes' physiology, set out in two posthumous works, the *Treatise on Man* and the *Description of the Human Body* (Descartes 1664). In these immensely influential works, Descartes likens the body to a machine and adopts Galen's division of the body into parts with specific functions and uses. However, these parts solely function according to the laws of mechanics and not natural faculties. Cartesian biology had its supporters, but it also provoked opposition from various schools of thought that employed different strategies of recomposition and theorization.

The iatromechanical trend derives from the speculation and experimental study of the spatial mobilization of anatomical forms. It seeks to explain the biological movement of living bodies in the same way that 17th-century rational mechanics attempted to understand inanimate bodies, following the example of Galileo, Descartes and Harvey. The iatromechanical anatomist considers the "geometric form" of reality and the "force" (*vis*) by which the form moves. The forces by which iatromechanics operates are essentially external to the structure they move (*impulsio*). The iatromechanical scholar, therefore, has several methods at his disposal: the model, an ideal geometric representation of the biological process to be analyzed; the experiment, which determines the scientific legitimacy of this a priori representation; and measurement, which is essential if one wishes to give a rational, mathematical expression to the experiment. The term for knowledge is "natural law", not conceived as a pure mathematical relationship between the data of measurement but as a general outline of the geometrical–dynamic process to which vital movement is confined. This is the inspiration for the speculative and experimental physiological essays of Borelli (1680), Baglivi (1704) and Boerhaave (1713).

2 For descriptions of some of these experiences, see Cunningham (2010).

The iatrochemical trend, meanwhile, stems from Paracelsus' (1589) experimental and quantitative rationalization of biology, leading to the iatrochemical study of vital processes. This is an attempt to understand the biological movement of living bodies by reducing it to the chemical processes occurring within the organism. The "elementary form" here corresponds to the substantial transformations of living matter; the "elementary chemical process" called *fermentatio*, which is a term derived from Paracelsus' "ferment", is a broad expression meaning "material transformation of a chemical nature". The "force" by which the chemical operation is generated or carried out is inherent in the matter that transforms and moves. It is no longer a simple mechanical *impulsio*, but an animating principle, a principle of life (*archeus*). Fermentation is the key to the transformation of substances, or from health to disease (Willis 1659). The iatrochemical scholar uses the methods given by Paracelsus for his research, namely, chemical experimentation (*experientia*) and the "sympathetic contemplation" of experimental reality. Boyle (1661) also added measurement to the initial iatrochemical experiments of van Helmont (1648) and Franciscus de le Boë, known as Sylvius (de le Boë 1681).

The modern era is characterized, in short, by the gradual reconfiguration of physiology from the concept of "movement of substances" to the concept of a set of natural laws. Disagreements then arose over the nature of these laws and knowledge of the body, which was either interpreted in the sense of a pure mechanism, a strategy of recomposition adopted by iatromechanics, or toward a mechanistic sense of the course of fermentation, a strategy adopted by iatrochemists, who stripped Paracelsus' alchemy of all its esoteric symbolism. Nevertheless, at the end of the 17th century, iatrophysics and iatrochemistry tended to reduce physiology to a branch of physics or chemistry.

In response to this trend and this conception of living matter and its overly "substantial" movements, "transmechanical" conceptions of living phenomena developed, represented by the doctrines of vitalism. This time, the aim was to determine specific concepts of life and, if possible, to formulate the laws governing what would come to be known as "biology" at the end of the century. The living matter of the body possesses a specific property, or "vital force", which disappears with death and enables it to respond specifically to external stimuli (reactive vitalism) through movement, sensation or both. By acting on inorganic matter, the vital force is also capable of spontaneously producing biological forms adapted to the external situation, or, more or less, new forms (configurative vitalism). This perspective was followed by biologists and physicians and became dominant during the Enlightenment in various forms, depending on the author. The vitalist claim for the autonomy of physiology led to research into living beings down new paths, such as the systematic exploration of the irritability and sensibility of fibers by von Haller (1752) and the study of tissues by Bichat (1799).

3.3.3. *Instrumentation*

We will later see how the theoretical inspiration of physiologists and the living systems they advocate are expressed in their approach to different physiological functions, as well as their position in controversies related to experimentation. However, alongside these systems, there is also physiological empiricism, corresponding to research carried out on animal life, less with a concern to interpret it than out of simple curiosity to know the facts, or with the intention of perfecting the experimentation and method.

The major contribution of Baroque-era physiologists was the introduction of measurement into physiological experimentation. Instrumentation also became one of the limiting factors in the young field of physiology. In this context, we can look at the symbolic contribution of Santorio Santori, a disciple of Girolamo Fabrizi d'Acquapendente. In 1629, Santori measured perspiration (water loss through the skin and lungs without apparent sweating) by weighing himself regularly using a scale of his own invention before and after meals, after exercise, at rest, during sleep, etc., to assess what was absorbed (food, drink), lost (excretion) or perspired. This is one of the first examples of a measuring instrument used for a planned biological experiment, which we call a "metabolic study". As a creative inventor, he developed other instruments: an anemometer, a hygrometer and, inspired by Galileo, a thermoscope that could detect the difference between hot and cold, which he transformed into an air thermometer. The thermometer thus became an instrument for measuring the temperature of living beings, enabling Santori (1690) to demonstrate the consistency of body temperature in mammals and to theorize thermal regulation. Animal heat could no longer be of cardiac origin, since Giovanni Borelli refuted Descartes' thermomechanical thesis using a thermometer, showing that the heart is at the same temperature as the animal's body. An "informed" observation was also made possible with the introduction of the microscope, paving the way for microscopic experimentation. Physics and chemistry laboratories were finally equipped with instruments capable of studying metabolic parameters (gas collection systems, gasometers, air pumps and pneumatic machines capable of creating a vacuum, calorimeters, precision scales) or excitation through the production of electricity (Leyden jar), which would benefit physiology.

3.3.4. *New methods of exploration*

In the 17th century, the philosophies of experience developed by Descartes (1931, 1637), Bacon (1620) and Boyle (1661, 1669) supported the idea of the necessity for strict protocols. In the context of specific scientific controversies, critical thinking contributes to the search for scientific rigor. It is a factor of progress, and this is one aspect of the usefulness of controversies, especially if they are contested within new institutions under the scrutiny of the community.

Bacon, Descartes and Boyle's ideas played a key role in the formation of learned societies and academies throughout Europe. This critical thinking, disseminated through scientific journals and encyclopedias, should, in theory, have been applicable to the life sciences and become more than just a general epistemological framework for the new science. This transposition had to be carried out by practitioners in these fields, who were dedicated to rigor in experimental work. As Diderot explained, "Observation of nature must be assiduous, reflection must be profound, and experience must be exact. These qualities are rarely found together. Creative geniuses are therefore rare" (Diderot 1754, p. 17). In addition to the accounts of experiments themselves, a review of 18th-century medical and naturalist literature reveals a few attempts to codify experience and define a specific method for studying living beings. The article *Observation* from the Encyclopedia, written by the physician Ménuret de Chambaud (1765), the various editions of Johann Georg Zimmermann's *Traité de l'expérience en général et en particulier dans l'art de guérir* (Zimmermann 1774) and Jean Senebier's *L'art d'observer* (Senebier 1775), represent some of these attempts, whose influence on actual practice should be better assessed. They at least demonstrate a reflective concern in this new field and would come to be known as "biology" in the following century, a complex science that would continue to grow.

3.4. The exploration of animal functions and experimental controversies

3.4.1. *Nutrition and digestion*

Paracelsus assumed the existence of a major archeus, which secondary archei specific to each organ controlling vital processes depended on. Van Helmont believed that, in addition to these archei, there was a chemistry specific to living beings, perfectly illustrated by the phenomenon of digestion, the various stages of which he described as follows: the transformation of food into chyle, chyle into blood, and then blood combined with air to produce vital spirits. For iatrochemists such as Sylvius, digestion was the preferred field of application for fermentation.

The digestive glands were explored. Reinier de Graaf used experimental fistulas and collected pancreatic secretions, whereas Marcello Malpighi recognized that the lobular structure of the liver meant that it was also a secretory gland. A significant step forward was taken when Astruc (1714) demonstrated the dissolving action of saliva, bile and pancreatic juice on food. On the other hand, for iatromechanists such as Borelli, digestion remained the breakdown of food into smaller and smaller particles under the effect of stomach contractions. This was also the theory of Boerhaave (1713), who described the transformation of food under the purely mechanical (and not chemical) action of the organs, starting with the teeth and chewing.

Digestion is also described as a mixed process. According to Georg Ernst Stahl, matter is corruptible; the role of nutrition is to replace corrupted matter, and that of excretion is to eliminate it. Apart from appetite, the buccal and stomach phases of nutrition can be understood mechanically. However, there is also digestion, which transforms food using enzymes to make it suitable for the body. Finally, nutrients are assimilated. Nutrition not only replaces corrupted matter but also enables the body to grow when intake exceeds excretion (Stahl 1708). The idea of a cycle with a constant renewal of matter, and that of the corruption of the body providing an outlet for ingested food, is relatively new. For Stahl, all these processes (appetite, digestion, distribution, assimilation and excretion) are directly dependent on the soul.

The debate on digestion resumed with the experiments of René-Antoine Ferchault de Réaumur and Lazzaro Spallanzani. Réaumur used small brass or lead tubes that were open at both ends and contained a piece of meat, which he fed to different species of birds. After a few hours of digestion, the meat was recovered through regurgitation. Réaumur (1752) demonstrated the chemical nature of digestive processes in vivo. Spallanzani then repeated the experiments, varying parameters such as the nature of the food, the temperature and the nature of the stomach, in other words, the species, including himself (self-experimentation). He also demonstrated the possibility of artificial digestion in vitro, without the intervention of mechanical agents or vital forces, by recovering gastric juice from birds (Spallanzani 1783). These processes were repeated in the 19th century to demonstrate the complexity of the phenomenon of digestion at different levels of the digestive tract using fistulas and cell homogenates.

3.4.2. *Relationship functions*

Nicolas Sténon's famous *Discourse on the Anatomy of the Brain* agrees with this: while the strategy of gradually breaking down the brain and the entire nervous system through anatomy has met with some success, the results of reassembling the parts and attempts by physiology to assign specific functions to them remain very poor (Sténon 1669).

However, the physiological spirit of the time was particularly evident in numerous studies on the nervous system. Many cerebral formations have been incorporated and systematized by anatomists, such as Thomas Willis, Félix Vicq d'Azyr, Domenico Cotugno, Samuel Thomas Soemmerring, Alexander Monro Secundus and Vincenzo Malacarne. However, anatomy in general was marked by a utilitarian spirit through systematization in the form of numerous treatises and the development of a remarkable topographical anatomy supporting the prospect of successful "anatomically regulated" surgical procedures. The keen curiosity of the nervous system over time can be explained by the fact that it seems to be the organ most directly responsible for what makes humans unique. Here, we consider a few questions related to its function, concerning nervous fluid, reflexes, brain locations and sense organs.

The Cartesian physiology of animal spirits, which had replaced Galen's theory of psychic pneuma, was just as speculative as the latter. Animal spirits were responsible for capturing sensations from the outside world and from within the body, transmitting these impressions to the brain and triggering muscle movements on the basis of the impressions received. Jan Swammerdam attempted to refute this, experimentally speaking. In his *Biblia naturae*, the chapter entitled "Experiments on a particular movement in the muscles of the frog, applicable to all movements of man and animals" describes experiments using the sciatic nerve and the gastrocnemius muscle of a frog. The 1658 experiment (demonstrated to the Duke of Tuscany) consisted of pinching the end of the nerve and attempting to feel the contraction of the muscle with both hands. Contraction did occur. Therefore, animal spirits do not necessarily come from the centers of the brain (ventricles). The 1664 experiment uses the gastrocnemius nerve–muscle preparation in a special device, where the orifice through which the nerve passes is sealed with a mixture of ichthyocol and starch and where a graduated glass tube allows the variations in muscle volume to be measured. The result was that the muscle did not increase in volume when contracted, whereas the supposed animal spirits should have caused it to swell. Francis Glisson subsequently confirmed that the volume of the arm does not increase through voluntary contraction of the muscle. The irritability of muscle fibers is sufficient to explain their contraction without being inflated by animal spirits, a contraction that can occur independently of the rest of the organism (Glisson 1677). However, Swammerdam's (1738) findings remained largely unknown because of the late publication of *Biblia naturae* and its English translation in 1758.

At the end of the 17th century, neuromuscular physiology also encountered opposition between iatromechanists and iatrochemists. Borelli saw the movements of the arms and legs as a simple mechanism on the basis of the principle of leverage. While according to Descartes, animal spirits produced by the brain spread mechanically through the nerves, Willis (1670) likened this propagation to the burning of a wick, followed by an explosion in the contracting muscle, an analogy more in line with the views of iatrochemists.

Throughout the 18th century, there was considerable confusion about the nature and existence of these animal spirits. The vitalist Albrecht von Haller preferred to use the term "nervous force" (*vis nervosa*) and, following Glisson, distinguished between the vital properties of muscle irritability and nerve sensibility: muscle activity is specific and can be autonomous, given that direct muscle excitation is possible. Von Haller reviewed the various theories of nervous force and its means of transport, which, according to the authors, included flammable or explosive gases, acids, sulfur or phosphorus compounds, or more subtle substances such as fire, light, Newton's ether and electricity (von Haller 1757–1766). The most common generic terms are "nerve juice" or "nerve fluid", the latter being the subject of an article by von Haller in *Supplément à l'Encyclopédie*, replacing animal spirits, which seem to have occult qualities (von Haller 1776). While there were remaining alternative mechanistic theories to animal spirits involving vibrations (Hartley 1749) or fiber movements (Bonnet 1754), certain experiments were designed to test the material reality of the concept of nerve fluid. Those who rejected the theory claimed that if it existed, ligating the nerve should have caused swelling of the nerve upstream, which was not the case. Supporters of this theory in favor of the fluid claimed that motor or sensory paralysis could be observed following ligation and that swelling did not occur because the juice or fluid was too fine a substance to cause it. Neurochemistry was still too limited to formulate anything other than hypotheses, despite the work of Giuseppe Francesco Borri and Johann Thomas Hensing and, in Lavoisiers' time, the memoirs of Fourcroy (1793).

The question of nerve fluid was taken up again in Italy through experiments on animal electricity. Galvani's (1791) treatise *De viribus electricitatis in motu muscularis commentarius* was the result of numerous experiments conducted between 1781 and 1791 on the physiological action of electric shocks in the form of natural atmospheric discharges or artificial electricity produced by machines (the Leyden jar, invented in 1746). The famous "galvanoscopic leg" experiment, using the sciatic nerve and gastrocnemius muscle, consists of touching the nerve and muscle with two metals joined at either end to cause muscle contraction. According to Galvani, muscle fibers are analogous to small Leyden jars, where the conductor connected to the internal armature is the nerve. When contact is established between the nerve and the surface of the muscle using the metal arc, the muscle fibers are discharged, causing contraction. According to him, this is proof of the existence of sources of electricity in living tissue. Nerve fluid is animal electricity that is produced naturally by the brain. Memory aroused considerable interest, as evidenced by the numerous articles and books on animal electricity that appeared at the time, including some spectacular works such as those by Aldini (1804). The vitalist sensibility of the time was well suited to work on galvanism, which made electricity the very essence of life by assimilating it to the life force; for this reason, the famous controversy with the physicist Alessandro Volta began. Volta repeated Galvani's experiments and reported that contractions also occurred when the ends of the metal arc touched only the nerves, leading him to reject the idea of the Leyden jar muscle. Interested in the bimetallic arc, he believed that it was the contact between metals that generated electricity, not fabrics: it was "metallic" electricity, not "animal" electricity. He therefore created the concept of electrical force (EMF), which replaces vital force (Volta 1792). A series of experiments and counterexperiments by the two protagonists followed. Convinced of the existence of animal electricity in nature, as evidenced by the case of the torpedo he had studied, Galvani sought to demonstrate the contraction of the galvanoscopic leg in the absence of any metal while refuting Volta's interpretations of his experiments with the bimetallic arc. Volta verified his hypothesis of electricity production through heterogeneous metal contacts, using a battery that would eventually carry his name, the first source of direct current in the history of physics. There was much controversy in the early 19th century, which saw the rise of new disciplines: electrical physics, electrochemistry and electromagnetism. However, Galvani's idea of electricity production through living tissue was unfairly discredited for a time, until new experiments were conducted by the Italian school (Leopoldo Nobili, Carlo Matteucci), and more importantly, quantitative studies were carried out by the German School of Electrophysiology.

The physiology of reflexes also generated great interest in the classical age. Involuntary movements have been puzzling since antiquity, as they appear to signal the disappearance of the subjective aspects of being. Descartes developed a theory of elementary movement, according to which there is no physiological mechanism that is identical to afferent movement (centripetal nerves) and efferent motor movement (centrifugal nerves): incident movement corresponds to traction in which nerves, likened to strings, transmit to the brain, causing pores in the cerebral cavity to open, allowing animal spirits to enter the nerves. Reflex movement corresponds to the movement of animal spirits that swell the muscles, with the nerves being likened this time to organ pipes. This heterogeneity of mechanisms is one of the reasons why Canguilhem (1955) rejected Descartes as the father of the concept of reflex, instead favoring Willis, who suggests a dual movement of transmission of animal spirits, with them being centrifugal and centripetal, likened to the combustion of an "inflammable substance". Descartes' theoretical physiological description was also challenged on experimental grounds by Swammerdam, Glisson and von Haller, as we have previously described. Sensorimotor functions were explored in the 18th century, using frogs as the preferred animal model, with several significant contributors. Jean Astruc acknowledged the existence of intracerebral obstacles (reflexes). However, movements were observed in decapitated animals. The brain is not the seat of, nor does it control, all motor functions (Astruc 1723). Robert Whytt was one of the first to provide experimental evidence in favor of the role of the spinal cord. In a decapitated frog, reflex movements persist. Stimulation of the spinal cord causes flexion of the lower limbs. In a decapitated frog in which the spinal cord is destroyed, there are no reflex movements. His logical conclusion was that reflexes involve the spinal cord and not the brain but also, as a good animist, that the action of the soul extends throughout the nervous system (Whytt 1751). Johann August Unzer developed the idea of excitomotor phenomena caused by sensory impressions that are not conscious because they do not reach the brain but are transmitted directly to the spinal cord or the sympathetic center and then reflected back to the muscle (Unzer 1771). Georges Prochaska also likens reflexes to behavior observed in response to stimulation, similar to a beam of light reflected off a mirror. The sensory nerves connected to the brain and the motor nerves connected to the muscle can, through their connection to the spinal cord, transform an impression into movement. This reflex depends on a *vis nervosa*, a term borrowed from von Haller, which is a force that is inherent in all nerve fibers connected to the medulla oblongata and the spinal cord. This is proven by the persistence of movement excitability below the spinal cord section in frogs. He also suggested the essential idea that reflexes participate in the principle of conservation of life, for example, as a defensive reaction to painful stimuli (Prochaska 1784). The concept of the reflex was therefore well defined by the end of the 18th century.

The physiology of the brain itself, however, remains unclear. The study of reflexes allows us to circumvent the delicate physiological problem of the nature of voluntary movements, which are supposed to originate in the brain, an organ that seems far more complex and much less accessible to experimentation than the medulla. The question of the location of brain functions is very old. Moving away from the ventricular interpretation of faculties by the Church Fathers, Descartes (1664) attributed the physiological role to animal spirits traveling through the cerebral substance, without precisely locating any faculties, with the pineal gland acting as an interface between the soul and the body. Willis (1664) made some proposals regarding the localization of functions in different parts of the brain. After him, the issue became clearer. François Gigot de Lapeyronie explained that if a sensation is caused by the flow of animal spirits in a specific part of the brain, damage to that part should deprive the soul of the sensation and that if some parts of the brain are damaged without any deterioration in mental activity being observed, this could be because mental activities had been taken over by other parts of the brain. He believed that, through clinical facts and autopsy data, he had proven that not all parts of the brain are essential to life and that not all parts are essential to the functions of the soul (La Peyronie 1741). In fact, only comparative anatomy and pathological anatomy allow the issue of localization to be addressed indirectly, and a few hypotheses need to be formulated. The issue is certainly physiological, but the epicenter of the debate is also philosophical, touching on free will and the materiality of the soul. Did proposing mechanisms for voluntary movements and other functions exclude free will? At the end of the 18th century, Soemmerring (1796) still titled his major anatomical work *Ueber das Organ der Seele* (On the Organ of the Soul), which drew criticism from Immanuel Kant.

While the question of brain localization remained unanswered by experiment (it was not until the 19th century that progress was made in this field), the physiology of the senses underwent significant advances. The empiricism and sensualism of the Enlightenment sharpened curiosity about the function of perception. The sense organs, with the nerves that connect them to the brain, are the source of all knowledge of the outside world. How are the properties of the outside world perceived by the senses conveyed to the brain, and to which specific area of the brain? The answers to these difficult questions lie in the individualization, for each sense organ, of the structure that constitutes its main sensitive part, the innervation of these organs and the sensory areas of the brain where the fibers end. These questions were addressed jointly, especially from the 18th century onward, in specialized works dealing with one or more of the sense organs concerning touch, taste, smell and hearing. However, it is vision, the only sense that the research methods specific to the physical and mathematical sciences could be applied to, that holds the most advanced research. For this reason, it interested almost all the great physicists of the time: Kepler, Descartes, Newton, Mariotte, etc. This concept also constitutes a broader field of physiological research than the other senses do, with a focus on the retina as the central organ of vision, the accommodation of the eye, color vision and the learning of vision. We need only mention the memoirs of François Pourfour du Petit, who, using freezing and sectioning of the eyes, showed that the lens is not at the center of the eyeball and therefore not the main organ of vision and who succeeded in assessing the quantity of humors in the anterior and posterior chambers (Petit 1723, 1726, 1728). Finally, Cheselden (1729), an English ophthalmologist, restored sight to a 14-year-old boy who had been blind since birth due to cataracts, experimenting with the origin of visual sensations. The physiology of vision then fueled the philosophical debate between innatists and empiricists on the famous Molyneux question.

3.4.3. *Reproductive functions*

The natural movement of a living form can mean a spatial displacement of that form and its functional consequences, the study of which is physiology. It can also be an ontogenetic process through which the form becomes what it is, this time being the object of embryology.

Until the mid-17th century, "liquid" epigenesis, used since antiquity, prevailed. The embryo was formed by the mixing of two liquid seeds, which had no defined organization. Various solid organs were formed successively by coagulation, in order of their presumed physiological importance. Epigenesis, the gradual construction of the embryo from a primitive, unorganized substance into an organized being, was made possible by a vegetative soul (Aristotle) or naturally generative and configurative faculties (Galen), entities that Descartes sought to dismiss by advocating strictly mechanistic epigenesis.

At the end of the 17th century, various discoveries overturned the liquid hypothesis. The question of generation was developed on the basis of three types of data: new anatomical observations concerning the reproductive system, microscopic anatomical data highlighting the elements depicted (micrographs) and certain experimental data.

The observation of embryonic development was limited to chicken eggs for a long period of time until Harvey (1651), a supporter of epigenesis who still believed in immaterial fertilization, made the first observations in mammals. According to Harvey, all living beings have the same mode of generation and originate from eggs. The experiments of Italian naturalist and physician Francesco Redi then demonstrated that maggots do not spontaneously generate from decaying meat but require fly eggs. Matter does not spontaneously take shape from unorganized matter (a mixture of liquid seeds) but requires a "germ", a "figurative" element, in other words, something endowed with structure, namely, the fly egg (Redi 1668). The observations of Niels Stensen (Sténon) on eggs in the "testicles" of female squalid fish (Sténon 1667) and various types of mammals (Sténon 1675a, 1675b), as well as Reinier de Graaf's quantitative observations and description of the ovarian follicles of female rabbits, which he mistook for eggs (de Graaf 1672), were also an important step in "figurism", as was Antoni van Leeuwenhoek's observation of spermatozoa (van Leeuwenhoeck 1677).

Following the experiments of Redi, the observations of Sténon, de Graaf and van Leeuwenhoek, generational theories were divided into "ovist" and "animalculist" versions, depending on whether the task of containing the preformed germ or the formless principle was attributed to eggs or animalcules. In doing so, by accrediting the idea that parents provide a "germ", a structure, or more simply a seminal fluid, and thus by postulating the pre-existence of structures, they called into question the epigenesis defended until then by Aristotle, Harvey and Descartes. The pre-existence of germs seemed to be congruent with Creation, whereas creationism seemed incompatible with spontaneous generation and epigenesis. It also established a convenient hypothesis that allowed certain biological mysteries to be circumvented: the determinism of organ formation, the development and progressive definition of physiological functions, and heredity.

This arbitrary elimination reduces physiology to the study of the mechanical functioning of organs and restricts its scope to the field of reproduction, but pre-existence requires solid evidence. The famous dispute between Caspar Friedrich Wolff, a supporter of epigenesis with newly formed intermediate embryonic forms (leaflets), and Albrecht von Haller, a physiologist who supported pre-existence, was therefore based on very detailed anatomical observations of chicken embryos. For von Haller, preformed structures are transparent and semifluid, which makes them invisible to us. Even if they are invisible, we cannot argue that they do not exist (von Haller 1758). For Wolff (1759, 1768, 1769), it is sufficient to note the non-existence of preformed structures and the existence of intermediate structures. A long controversy ensued, with epigenesis gaining traction only in the following century. This can be explained by the near impossibility of experimentation, the divergence in the interpretation of observations and the return of the difficult questions that had been circumvented by the pre-existence argument: the inability to conceive of a strictly mechanistic determinism for epigenesis, in particular, the simultaneous construction of physiological functions, as well as heredity.

From the Renaissance to the Enlightenment, the generational dispute shifted from traditional questions, such as the location of development and the role of different organs (matrix-ovary) and fetal nutrition, to the role of figurative elements in the origin of being: egg (ovism of Sténon and de Graaf) or animalcule (animalculism of van Leeuwenhoek). It also concerned the question of the type of development: the pre-existence of structures in the figurative elements with or without overlapping (Jan Swammerdam) and with or without dissemination (Nicolaas Hartsoeker); or, again, various types of epigenesis, such as mechanistic epigenesis (René Descartes), molecular epigenesis (Georges-Louis Leclerc de Buffon, Pierre Louis Moreau de Maupertuis) and vitalist epigenesis (Caspar Friedrich Wolff, John Tuberville Needham). The question of generation is therefore highly complex, given that combinations and moderately appeasing positions are also possible. It was on the basis of these rival explanatory systems – alongside which a prolific and imaginative medical literature developed (Darmon 1977) – that specific biological questions such as those of spontaneous generation and "monsters" were discussed, combining religious arguments, philosophical positions, observational data and experimental data. From the 17th century to the 18th century, embryology followed a development that was quite similar to that of physiology in that it moved from a conception of generation as a simple static development of pre-existing forms to a dynamic epigenesis, in which form was set in motion. However, in the modern era, this was reflected more in the rivalry between different explanatory systems than in the establishment of an experimental science.

However, the question of spontaneous generation, as taken up experimentally by John Tuberville Needham and Lazzaro Spallanzani, must be considered separately. In the 18th century, spontaneous generation shifted from small organisms to microscopic organisms. Needham's belief that such generation existed was based on experimental arguments and, to a large extent, on a vitalist conviction that nature possesses a creative capacity (vegetative force). He started with the observation of microorganisms appearing in various infusions of plants or meat, in other words, from decomposing organic matter, one of the most famous being the eelworms that appear in infusions of honeyed wheat, which earned him the nickname "eelmonger" from Voltaire. Needham sought to control the parameters of his observations. For example, to study the appearance of microorganisms in meat juice, Needham enclosed the latter in a very hot, hermetically sealed flask, which he then placed in hot ashes to destroy any germs or eggs that might be suspended in the residual air. From his various observations, Needham (1745, 1749) concluded that microorganisms appear by spontaneous generation and not by reproduction, similar to more complex living beings.

Needham did not, strictly speaking, develop a general physiology or even embryology, although he was a supporter of vitalist epigenesis. Lazare Spallanzani, on the other hand, was an ovist preformationist when he set out to demonstrate the lack of experimental rigor in Needham's work, which, in his view, rendered it scientifically worthless. He criticized Needham for not heating his preparation sufficiently, to which Needham replied that by heating it too much, Spallanzani had destroyed the vegetative force within it. Spallanzani then designed his own rigorous experiments, which led to completely different results. The animalcules in the infusions always came from germs carried by the air. Therefore, germs also pre-exist: there is no spontaneous generation for microscopic beings either (Spallanzani 1765). Spallanzani's work, which complemented that of Redi, was considered a model of experimental rigor, as was that of Louis Pasteur on the same subject. Spallanzani (1785) also distinguished himself through his experiments on fertilization in amphibians, demonstrating the need for physical contact between sperm and egg during fertilization. While this discovery and the achievement of various artificial fertilizations did not lead him to reject his ovist approach, it did prevent him from understanding that it is the spermatozoa in semen that are fertile, even though he had discovered that semen deprived of spermatozoa by filtration was infertile, which he attributed to a loss through filtration of the stimulating fertile properties of the seminal fluid.

Abraham Trembley's experimental zoological work on regeneration is also worth mentioning. In his *Mémoires pour servir à l'histoire d'un genre de polype d'eau douce, à bras en forme de cornes* (Memoirs concerning the natural history of a type of freshwater polyp with arms shaped like horns), Trembley (1744) demonstrated the animal nature of the hydra and described a form of complete animal regeneration for the first time. Trembley's zoology was no longer purely descriptive: he systematically used experimentation to try to understand how animals functioned, introducing disruptions to understand the mechanisms underlying his observations (cross-sections, longitudinal fissuring and grafts). Trembley was thus the first to shed light on animal reproduction by budding, and Spallanzani showed that certain vertebrates also have this capacity for regeneration. According to Buffon, the same type of process must have been at work during these regenerations and during embryogenesis, with the child being conceived as the reproduction of its parents, which is understood as a type of regeneration of the entire organism. The term "reproduction", which was mainly used to refer to the regeneration of parts of an amputated organism at the time, subsequently became generalized in biology to mean the action of the physiological functions by which living beings perpetuate themselves, replacing the older term "generation", meaning the act of generating.

3.5. Chemical exploration of respiration

3.5.1. *Places of respiration and animal heat*

The location of respiratory function in the lung became clearer after the discovery of pulmonary circulation. Vesalius experimented with mechanical ventilation in pigs by blowing air through a reed inserted into the respiratory tract, demonstrating that ventilation was necessary for life. However, he believed that it was the ventilatory movements themselves that were necessary for life and that their cessation therefore caused death (Vesalius 1543). These experiments were repeated more than a century later by Robert Hooke, who also proposed the idea of gas exchange associated with ventilatory movements. Hooke revealed that if the thorax was opened, the lungs were pierced, and a stream of air was pushed through the trachea by bellows and escaped through the arterial openings on the surface of the lungs, life was maintained even if there were no ventilatory movements. These movements are therefore essential to life, but only insofar as they ensure the renewal of air in the lungs (Hooke 1667). However, Hooke did not address the question of whether respiration represents the entry of some substance into the blood or the exit of some element into the exhaled air.

Malpighi (1661) used microscopic and macroscopic approaches to show that air passes through the trachea and then descends to the bronchi in the pulmonary alveoli and described the pulmonary capillaries. It was then clear that air descended into these alveoli while blood flowed into them, and a physiological link between air and blood was therefore likely. However, Malpighi focused his attention on the mechanisms of air exchange, limiting the existence of air to these alveoli. Hooke confirmed that there was no air in the pulmonary vein and that air could not be transported to the LV, as Galen had claimed. It was around 1650 that the role of air was truly addressed, as we shall see later.

This is because the focus was still mainly on the question of heat, with attempts also being made to break free from the Galenic approach. For iatromechanists, respiration was sufficiently explained by the action of the intercostal muscles and the expansion of the lung mass, which served to dilate and expand the vessels of the lungs. The passage of blood through the fine pulmonary vessels is used to break down the blood, fluidize it by wearing down its largest particles and warm it. The role of the lungs is to warm the blood, not to cool it. The arterial blood that leaves the lungs therefore warms the heart, which is cooled by the venous blood coming from the rest of the body. These ideas about animal heat were broadly taken up, with variations, by Friedrich Hoffmann, Georg Ernst Stahl and Stephen Hales.

3.5.2. *Respiration and its link to blood and air chemistry*

Although it was not abandoned, the question of heat took a back seat in the mid-16th century, giving way to the mechanisms of pulmonary ventilation, which were studied, for example, in Oxford by Bathurst (1654). More fundamentally, however, iatrochemists sought to interpret respiration chemically and were interested in the seemingly related changes in the composition of blood during circulation and in the composition of the air we breathe. Richard Lower, a pioneer of blood transfusion, noted that dark-colored blood enters the lungs and comes out bright red, suggesting that something in the inhaled air mixes with the blood, changing its color (Lower 1669). Furthermore, Malpighi's discovery of the internal contact between air and blood in the pulmonary alveoli led to the idea that it was not air as a whole that was life-giving but rather a portion of the air in the blood. Is it the entry of a vital element or the expulsion of noxious substances that occur during respiration? Does something pass from the air into the blood, or does the blood release some of its contents? This physiological question, therefore, raises a chemical question: air is not simple but is composed of several parts. The composition of air then became the focus of chemists, ushering in the field of pneumatic chemistry.

This pneumatic chemistry began after Jean-Baptiste van Helmont, inventor of the word "gas", distinguished between different types of gases. Among these, *gas sylvestris* (carbon dioxide) results from the combustion of coal, the effect of vinegar on certain stones or the fermentation of grape juice. In Oxford, Robert Boyle and John Mayow manipulated these gases using glass bells. Later, Stephen Hales described a simple device consisting of a tube turned upside down over a container filled with water to collect the gases. According to Hales (1733), ordinary air is "elastic air", which, when used, loses its elasticity and becomes "fixed air". Joseph Black later demonstrated that Hales' fixed air (carbon dioxide) did not have the same properties as ordinary air, that it was present in exhaled air and that it was identical to van Helmont's *gas sylvestris* released during fermentation (Black 1756).

Chemists in Oxford had almost managed to decipher pulmonary hematosis. Using a pneumatic machine capable of creating a vacuum (an air pump), Robert Boyle observed that mammals cannot survive without air, owing, in his opinion, to the absence of a vital essence that animates the "vital spirits". Boyle suspected the existence of a vital substance responsible for combustion, respiration in living beings and fermentation of grape juice. Noting that these phenomena consume only a small amount of this substance, he concluded that air is a mixture (Boyle 1660). John Mayow demonstrated through a series of experiments that air and saltpetre contain a common component, "nitro-aerial spirit". There is a life-giving element in air that serves to maintain combustion and respiration, a process that Mayow (1674) considered analogous. Oxygen would not be discovered as that element until more than a century later.

3.5.3. *Respiration according to phlogiston theory*

The results of chemical analysis of air and experiments conducted by Carl Wilhem Scheele, Joseph Priestley and Henry Cavendish in the 1770s were interpreted in a complex manner in light of Georg Ernst Stahl's phlogiston theory, according to which a burning body loses a substance called "phlogiston". When Scheele and Priestley prepared "fire air" or "ignited air" (Lavoisier's oxygen) shortly before Lavoisier, they believed it to be a combination of phlogiston and water. Similarly, by preparing oxygen through the calcination of mercury oxide, Priestley believed that he had isolated "dephlogisticated air" (actually nitrous oxide), while "phlogisticated air" (nitrogen), released during combustion, could not sustain fire. The "mephitic air" or "asphyxiating air" (Lavoisier's nitrogen) of Scheele and Cavendish, which, unlike oxygen, does not sustain life, was named "phlogisticated air". When Cavendish described the preparation of hydrogen ("inflammable air"), he believed that he had identified phlogiston, etc.

Priestley compared fixed air (carbon dioxide) with dephlogisticated air, or pure or vital air (oxygen), and was the first to define its vital properties. A plant can release this air in a bell jar, keeping a candle lit. This air also allows a mouse to survive under the bell jar. His "experiments and observations on different types of air" relate to the change in color of blood when it comes into contact with different types of air, either dephlogisticated or fixed. The blood turns dark, similar to venous blood, when it comes into contact with fixed air or gas sylvestre (carbon dioxide), and it turns bright red, similar to arterial blood, in the presence of dephlogisticated air (oxygen). However, black blood enters the lungs, and bright red blood leaves them. Priestley subsequently developed respiratory physiology on the basis of phlogiston theory. Blood absorbs phlogiston introduced into the body through food, then enters the lungs loaded with phlogiston and leaves them dephlogisticated, with exhalation eliminating the excess phlogiston. During respiration, inhalation brings in dephlogisticated air and exhalation brings in phlogisticated air, making it unbreathable and simultaneously forming fixed air (Priestley 1774–1786).

Owing to phlogiston theory, chemical phenomena, such as combustion, and physiological phenomena, such as respiration, were given a common explanation. Pre-Lavoisian chemistry already maintained that respiration had a significant effect on the air in the atmosphere, changing its nature and causing it to lose its ability to sustain animal life. The respiratory phenomenon was correctly reduced to gas exchange, in the sense of exchanges of parts of air (gas) between the blood and the external environment. However, only one aspect of the respiratory phenomenon was perceived: the release of a substance (phlogistic), without the need for it to enter the lungs.

3.5.4. *Lavoisier, respiratory gas exchange and animal heat*

Antoine-Laurent de Lavoisier provided a completely different interpretation of these chemical and physiological phenomena. The experiments he conducted with Armand Seguin on respiratory gas exchange and with Pierre-Simon Laplace on animal heat, published between 1777 and 1790, considerably clarified the question of respiration to the point that he was considered the founder of nutrition and biological energetics. His chemical experiments on the calcination of mercury and the reduction of mercury oxide ("mercuric lime") led him to challenge the phlogiston theory and his experiments on living organisms and Priestley's theory of respiration, which was based on it.

Shortly after Scheele (ignited air) and Priestley (dephlogisticated air) performed their work, Lavoisier named this dephlogisticated air "highly breathable", "oxygen", believing that it was the constituent of all acids and that oxidation and acidification were related. However, he fully recognized the biological significance of the new gas. He demonstrated the chemical similarity between calcination, combustion and respiration; he measured the amount of oxygen consumed and the amount of carbon dioxide and water produced during respiration, and he provided the first true chemical explanation of pulmonary hematosis. According to him, in respiration, such as in combustion, the air in the atmosphere provides oxygen and the "caloric" and "fluid" characteristic of heat. During respiration, it is the very substance of the animal, or rather its blood loaded with food, that provides the fuel. According to Lavoisier, the lung is also a place where blood is heated, but his explanation is different: respiration is a combustion taking place in the lung itself (Lavoisier 1777; Seguin and Lavoisier 1789).

Alongside his work on respiratory gas exchange, and together with Pierre-Simon Laplace, Lavoisier approached the problem of heat production. The two men managed to measure the amount of heat released by combustion in a calorimeter kept at zero degrees by melting ice and weighing the amount of melted ice. By applying this calorimetric technique to the study of animal heat, they determined the amount of heat produced by guinea pigs enclosed in their devices. The heat released by the animals was measured, as well as the amount of carbon dioxide produced by their respiration. The fact that respiration and combustion produce the same amount of heat for the same amount of carbon dioxide released means that respiration is a continuous production of heat similar to slow combustion, which is transmitted to the blood passing through the lungs and thereby spreads throughout the animal. Furthermore, Lavoisier and Laplace demonstrated that the amount of heat produced was greater than that which should result from the amount of gas (carbonic acid) formed at the same time as respiration. This led to the additional hypothesis that, in addition to the combustion of carbon-producing carbon dioxide, there is also a combustion of hydrogen with the formation of water (Lavoisier and Laplace 1780). Lavoisier supplemented these hypotheses with a study with Seguin on the role of perspiration, which promotes the removal of excess heat and thus contributes to thermoregulation (Seguin and Lavoisier 1790).

An analysis of Lavoisier's work shows that his denial of phlogiston was gradual. By focusing on the entry of oxygen (highly breathable air), he came to a correct definition of gas exchange and eventually spoke of the release of harmful carbon dioxide in exhaled air, refuting Priestley's phlogiston theory. He mainly explained the production of heat with a new idea: the slow combustion of carbon but also of hydrogen from food. The physiological link between nutrition, respiration and heat production was thus established. Lavoisier, Seguin and Laplace proposed replacing the idea of respiration, which is often still interpreted in vitalistic terms, with a deterministic perspective that could be interpreted through physical and chemical concepts. In contemporary terms, respiration became a thermogenesis step that was necessary for homeostasis.

Despite certain unresolved issues, such as the location of combustion, which he stated as being in the lungs despite Spallanzani's work on tissue respiration, and his continued adherence to caloric theory, Lavoisier presented a coherent idea of the animal machine, the functioning of which he believed to be regulated: "The animal machine is mainly governed by three main regulators: respiration, which consumes hydrogen and carbon and provides heat; perspiration, which increases and decreases depending on whether more or less heat needs to be removed; and lastly, digestion, which returns what is lost through respiration and perspiration to the blood" (Seguin and Lavoisier 1789, p. 580). After Lavoisier's death, attempts were made to apply this model to humans. It was largely due to these experimental contributions that iatrochemistry became animal chemistry in the 19th century.

3.6. Conclusion

This brief review of the exploration of animal functions should not obscure the fact that plant physiology was not neglected, from the rudimentary experiments of van Helmont to the work on nutrition, growth and gas exchange in plants by Marcello Malpighi, Stephen Hales, Joseph Priestley, Jean Ingen-Housz and Jean Senebier, who wrote the first chapter on photosynthesis. Whether they concerned plants, animals or humans, the common thread running through these modern experiments was to test certain hypotheses concerning the structure, connections, properties and usefulness of the parts of organisms. We hope to have shown the error that would result from hastily minimizing their scientific and epistemological significance, given the erratic progress of contemporary biology.

In the medical field, however, physiology's autonomy from anatomy, which remains the dominant science, is still imperfect. Many object to the claims of a physiology that is overly rational or theoretical about causes, without the reassuring support of anatomy (Cunningham 2010, pp. 156–162; De Angelis 2013). Admittedly, the discipline taught at the end of the 18th century called "physiological anatomy" or "anatomical physiology", for example in the *Programmes des cours d'enseignement dans l'école de santé de Montpellier* (1795), was a science that no longer sought to be merely descriptive, as Vesalius' anatomy had been, but an interpretation of functions. However, the momentum of experimental research entirely dedicated to identifying efficient causes, which requires a combination of robust theorization, rigorous and systematic methods and powerful tools, would not fully characterize physiology until the 19th century.

3.7. References

Primary sources

d'Acquapendente, F. (1603). *De venarum ostiolis*. Lauretii Pasquati, Padua.

Al-Nafis, I. (1547). *Ebenefis philosophi... expositio super quintum canonem Avicennæ. Tractatus de ponderibus et mensuris*. Comino, Venice.

Aldini, G. (1804). *Essai théorique et expérimental sur le galvanisme, avec une série d'expériences faites en présence des commissaires de l'institut national de France, et en divers amphithéâtres de Londres. Avec Planches*. Fournier Fils, Paris.

Aselli, G. (1627). *De lactibus, sive lacteis venis*. Giovanni Battista Bidelli, Milan.

Astruc, J. (1714). *Traité de la cause de la digestion, où l'on réfute le nouveau sistème de la trituration & du broïement. Et où l'on prouve que les alimens sont digérez & convertis en chile, par une veritable fermentation*. Ant. Colomiez, Toulouse.

Astruc, J. (1723). De phantasia sive imaginatione. In *Disputationum anatomicarum selectarum. Volume 4. Sensus externi, interni, respiratio*, von Haller, A. (ed.). Apud Abram Vandenhoeck, Gottingæ.

Bacon, F. (1620). *Novum organum scientiarum*. Adrianum Wiingaerde et Franciscum Moiardum, Lugdunum Batavorum.

Baglivi, G. (1704). *Opera omnia medico-practica*. Claudium Rigaud, Paris.

Bartholin, T. (1652). *De lacteis thoracicis in homine, brutisque nuperrime inventis historia anatomica publice proposita*. Typis M. Martzam, Hafniae.

Bathurst, R. (1654). Praelectiones tres de respiratione. In *The Life and Literary Remains of Ralph Bathurst*, Warton, T. (ed.). R. and J. Dodsley, London.

Bichat, X. (1799). *Traité des membranes en général, et de diverses membranes en particulier*. Veuve Richard, Paris.

Black, J. (1756). Experiments upon magnesia alba, quicklime, and some other alcaline substances. In *Essays and Observations: Physical and Literary*, volume 2. Philosophical Society of Edinburgh, Edinburgh.

de le Boë, F.S. (1681). *Opera medica, hoc est disputationum medicarum decas, methodi medendi libri duo, ideae novae praxeos medicae libri tres, ad eosque appendix, variaque alia opuscula...* Apud Samuelem de Tournes, Geneva.

Boerhaave, H. (1713). *Institutiones medicæ, in usus annuæ exercitationis domesticos digestæ. Editio altera prima longe auctior*. J. van der Linden, Lugduni Batavorum.

Bonnet, C. (1754). *Essai de psychologie*. Elie Luzac, Leiden.

Borelli, G.A. (1680). *De motu animalium*. Angelo Bernabo, Rome.

Boyle, R. (1660). *New Experiments Physico-Mechanicall, Touching the Spring of the Air and Its Effects (Made, for the Most Part, in a New Pneumatical Engine)*. H. Hall, Oxford.

Boyle, R. (1661). *The Sceptical Chymist*. J. Cadwell, London.

Boyle, R. (1669). *Certain Physiological Essays and Other Tracts...* Henry Herringman, London.

Cesalpino, A. (1571). *Quaestionum peripateticarum libri quinque*. Iuntas, Venice.

Cheselden, W. (1729). An account of some observations made by a young gentleman who was born blind, or lost his sight so early, that he had no remembrance of ever having seen, and was couch'd between 13 and 14 years of age. *Philosophical Transactions*, 35, 447–452.

Colombo, R. (1559). *De re anatomica libri XV*. Nicolai Beuilacquae, Venice.

Descartes, R. (1637). *Discours de la méthode, pour bien conduire sa raison, & chercher la vérité dans les sciences, plus la dioptrique, les météores et la géométrie, qui sont des essais de cette méthode*. Jan Maire, Leiden.

Descartes, R. (1664). *L'Homme de René Descartes et un traité de la formation du fœtus du même auteur, avec les remarques de Louis de la Forge, docteur en Médecine, demeurant à la Flèche, Sur le Traité de L'Homme de René Descartes, & sur les Figures par lui inventées*. Charles Angot, Paris.

Descartes, R. (1931). *Regulae ad directionem ingenii*. Vrin, Paris.

Diderot, D. (1754). *Pensées sur l'interprétation de la nature*. Paris.

Dionis, P. (1690). *Anatomie de l'homme suivant la circulation du sang et les dernières découvertes*. Laurent d'Houry, Paris.

Fernel, J. (1554). *Medicina*. Andreas Wechel, Paris.

Fernel, J. (1655). *Les sept livres de la physiologie*. Jean Guignard le Jeune, Paris.

Fourcroy, A.F. (1793). Examen chimique du cerveau de plusieurs animaux. *Annales de Chimie*, 16, 282–322.

Galien, C. (1538). *De usu partium corporis humani Libri XVII*. Chrétien Wechel, Paris.

Galvani, L. (1791). *De viribus electricitatis in motu musculari commentarius*. Typographia Instituti Scientiarum, Bologna.

Glisson, F. (1677). *Tractatus de ventriculo et intestinis*. Jacobum Juniorem, Amsterdam.

de Graaf, R. (1668). *De virorum organis generationi inservientibus, de clysteribus et de usu siphoni in anatomia*. Ex Officina Hackiana, Lugduni Batavorum.

de Graaf, R. (1672). *De mulierum organis generationi inservientibus tractatus novus : demonstrans tam homines & animalia caetera omnia, quae vivipara dicuntur, haud minus quàm ovipara ab ovo originem ducere*. Ex Officina Hackiana, Lugduni Batavorum.

Hales, S. (1733). *Statical Essays Containing Haemastaticks*. Innys & Manby, London.

von Haller, A. (1752). De Partibus Corporis Humani sensilibus et irritabilibus commentationes duae. In *Commentarii Societatis Regiae Scientiarum Gottingensis*. Apud Viduam Abrami Vandenhoeckii, Göttingen.

von Haller, A. (1757–1766). *Elementa physiologiæ corporis humani*, 8 volumes. Sumptibus M. M. Bousquet et Sociorum, Lausanne.

von Haller, A. (1758). *Sur la formation du cœur dans le poulet ; sur l'œil ; sur la structure du jaune*. M.-M. Bousquet, Lausanne.

von Haller, A. (1776). Fluide nerveux. In *Supplément à l'Encyclopédie*, volume 3, Panckoucke, C.-J. and Robinet, J.-B. (eds). Rey, Amsterdam.

Hartley, D. (1749). *Observations on Man, His Frame, His Duty, and His Expectations*. J. Leake/W. Frederick, London.

Harvey, W. (1628). *Exercitatio anatomica de motu cordis et sanguinis in animalibus*. William Fitzer, Frankfurt.

Harvey, W. (1651). *Exercitationes de generatione animalium*. Typis Du-Guardianis, London.

van Helmont, J.B. (1648). *Ortus medicinae, id est Initia physicae inaudita. Progressus medicinae novus, in morborum ultionem, ad vitam longam*. Apud Ludovicum Elzevirium, Amsterdam.

Hooke, R. (1667). An account of an experiment made by Mr. Hook, of preserving animals alive by blowing through their lungs with bellows. *Philosophical Transactions*, 2, 539–540.

Hunter, J. (1794). *A Treatise on the Blood, Inflammation, and Gun-shot Wounds*. George Nicol, London.

La Peyronie, F.G. (1741). Observations par lesquelles on tâche de découvrir le véritable lieu du cerveau dans lequel l'âme exerce ses fonctions. *Histoire de l'Académie royale des sciences*, 199–218.

Lavoisier, A.L. (1777). Expériences sur la respiration des animaux et dans les changements qui arrivent à l'air en passant par leurs poumons. *Histoire de l'Académie royale des sciences*, 185–194.

Lavoisier, A.L. and Laplace, P.S. (1780). Mémoire sur la chaleur. *Histoire de l'Académie royale des sciences*, 355–408.

van Leeuwenhoek, A. (1677). Letter 22 of 1677-11-09 (AB 35) to William Brouncker [Online]. Available at: https://lensonleeuwenhoek.net/content/wrote-letter-22-1677-11-09-ab-35-william-brouncker.

Lower, R. (1669). *Tractatus de corde, item de motu & calore sanguinis et chyli in eum transitu.* D. Elzevirium, Amsterdam.

Malpighi, M. (1661). *De pulmonibus observationes anatomicae.* Giovanni Battista Ferroni, Bologna.

Mayow, J. (1674). *Tractatus quinque medico-physici.* Sheldonian Theatre, Oxford.

Ménuret De Chambaud, J.J. (1765). Article observation. In *Encyclopédie, ou Dictionnaire raisonné des sciences, des arts et des métiers*, volume 11, Diderot, D. and D'Alembert, J. (eds). S. Faulche et Cie, Neuchâtel.

Monro Secundus, A. (1785). *The Structure and Physiology of Fishes Explained and Compared with Those of Man and Other Animals.* C. Elliot, Edinburgh.

Needham, J.T. (1745). *An Account of Some New Microscopical Discoveries.* F. Needham, London.

Needham, J.T. (1749). *Observations upon the Generation, Composition and Decomposition of Animal and Vegetable Substances.* London.

Nuck, A. (1691). *Adenographia curiosa et uteri foeminei anatome nova.* Samuel Luchtmans, Leiden.

Paracelsus (1589). *Theil Der Bücher und Schrifften, des Edlen, Hochgelehrten und Bewehrten Philosophi und Medici, Philippi Theophrasti Bombast von Hohenheim.* Conrad Waldfirch, Basel.

Pecquet, J. (1651). *Experimenta nova anatomica, quibus incognitum hactenus chyli receptaculum, et ab eo per thoracem in ramos usque subclavios vasa lactea deteguntur.* S. et G. Cramoisy, Paris.

Peiresc, N.C.F. (1634). Lettre de Peiresc à Dupuy du 25 juillet 1634 ; Lettre de Peiresc à Pierre Bourdelot du 6 septembre 1634. In *Lettres de Peiresc (1888-1898)*, Tamizey, P. (ed.). Imprimerie Nationale, Paris.

du Petit, F.P. (1723). Mémoire sur les yeux gelés, dans lequel on détermine la grandeur des chambres qui renferment l'humeur aqueuse. *Histoire de l'Académie royale des sciences*, 38–55.

du Petit, F.P. (1726). Mémoire sur plusieurs découvertes faites dans les yeux de l'homme, des animaux à quatre pieds, des oiseaux et des poissons. *Histoire de l'Académie royale des sciences*, 69–83.

du Petit, F.P. (1728). Différentes manières de connoître la grandeur des chambres de l'humeur aqueuse dans les yeux de l'homme. *Histoire de l'Académie royale des sciences*, 289–300.

Priestley, J. (1774-1786). *Experiments and Observations on Different Kinds of Air*, 6 volumes. J. Johnson, London.

Prochaska, G. (1784). *De functionibus systematis nervosa.* Wolfgang Gerle, Prague.

de Réaumur, R.A.F. (1752). Sur la digestion des Oiseaux. Premier mémoire. Expériences sur la manière dont se fait la digestion dans les Oiseaux qui vivent principalement de graines et d'herbes, et dont l'estomac est un gésier. Deuxième mémoire. De la manière dont elle se fait dans l'estomac des Oiseaux de proie. *Histoire de l'Académie royale des sciences*, 266–307/461–495.

Redi, F. (1668). *Esperienze intorno alla generazione degl'insetti*. All'insegna della Stella, Florence.

Riolan, J. (1649). *Opuscula anatomica nova : quæ nunc primùm in lucem prodeunt : instauratio magna physicæ & medicinæ per novam doctrinam de motu circulatorio sanguinis in corde : accessere notæ in Joannis Wallæi duas Epistolas de circulatio-ne sanguinis*. Typis Milonis Flesher, London.

Rudbeck, O. (1653). *Nova exercitatio anatomica exhibens ductus hepaticos aquosos & vasa glandularum serosa*. Eucharius Lauringerus, Arosa.

Ruysch, F. (1665). *Dilucidacio valvorum in vasis lymphaticus et lacteis*. Harmani Gael, The Hague.

Santori, S. (1690). *De statica medicina et de responsione ad staticomasticem*. Sumptibus Antonii Cellier, Lyon.

Seguin, A. and Lavoisier, A.L. (1789). Premier mémoire sur la respiration des animaux. *Histoire de l'Académie des sciences*, 566–584.

Seguin, A. and Lavoisier, A.L. (1790). Premier mémoire sur la transpiration des animaux. *Histoire de l'Académie des sciences*, 601–612.

Senebier, J. (1775). *L'art d'observer*. C. Philibert & B. Chirol, Geneva.

Servetus, M. (1553). *Christianismi restitutio. Totius ecclesiae apostolicae est ad sua limina vocatio*. Arnollat, Vienna.

Soemmerring, S.T. (1796). *Ueber das Organ der Seele*. Nicolovius, Königsberg.

Spallanzani, L. (1765). *Saggio di osservazioni microscopiche concernenti il sistema della generazione dei Seignori di Needham e Buffon. Dissertazioni due dell'Abate Spallanzani*. Per gli Eredi di Bartolomeo Soliani, Modena.

Spallanzani, L. (1783). *Expériences sur la digestion de l'homme et de différentes espèces d'animaux*. Barthelemi Chirol, Geneva.

Spallanzani, L. (1785). *Expériences pour servir à l'histoire de la génération des animaux et des plantes. Avec une ébauche de l'Histoire des êtres organisés avant leur fécondation par Jean Senebier*. Chez Barthelemi Chirol, Geneva.

Stahl, G.E. (1708). *Theoria medica vera, physiologiam et pathologiam, tanquam doctrinæ medicæ partes vere contemplativas, e naturæ et artis veris fundamentis sistens*. Litteris orphanotrophei, Halæ.

Sténon, N. (1664). *De musculis et glandulis observationem specimen cum epistolis duabus anatomicis*. Matthias Godicchen, Copenhagen.

Sténon, N. (1667). *Canis carchariae dissectum caput. Dissectus piscis ex canum genere. Appendix to Elementorum myologiae specimen.* Stellae, Florence.

Sténon, N. (1669). *Discours de Monsieur Sténon sur l'anatomie du cerveau à Messieurs de l'Assemblée qui se fait chez Monsieur Thévenot.* Robert de Ninville, Paris.

Sténon, N. (1675a). Observationes anatomicae spectantes ova viviparorum. *Acta Hafniensia*, 2(88), 210–218.

Sténon, N. (1675b). Ova viviparorum spectantes observationes. *Acta Hafniensia,* 2(89), 19–232.

Swammerdam, J. (1737). *Bybel der natuure*, volume 1. Isaak Severinus, Leiden.

Swammerdam, J. (1738). *Biblia naturae*, volume 2. Isaak Severinus, Leiden.

Trembley, A. (1744). *Mémoires pour servir à l'histoire d'un genre de polype d'eau douce, à bras en forme de cornes.* Durand, Paris.

Unzer, J.A. (1771). *Erste Gründe einer Physiologie der eigentlichen thierischen. Natur thierischer Körper.* Weidmann, Leipzig.

Valverde, J. (1556). *Historia de la composicion del cuerpo humano.* Antonio Salamanca & Antonio Lafrery, Rome.

Vesalius, A. (1543). *De humani corporis fabrica libri septem.* Johann Oporinus, Basel.

Volta, A. (1792). *Memorie sull elettricitá animale.* Presso Baldassare Comini, Pavie.

Whytt, R. (1751). Of the share which the mind has in producing the vital and other involuntary motions of animals. In *An Essay on the Vital and Other Involuntary Motions of Animals*. Hamilton, Balfour and Neill, Edinburgh.

Willis, T. (1659). *Diatribae duae medico-philosophicae (De fermentatione, De febribus, De urinis).* Apud Adrianum Viacq, The Hague.

Willis, T. (1664). *Cerebri anatome: cui assessit nervosum description et usu.* J. Martyn and J. Allestry, London.

Willis, T. (1670). *De motu musculari.* J. Martyn and J. Allestry, London.

Wolff, C.F. (1759). *Theoria generationis.* Hendel, Halle.

Wolff, C.F. (1768). De formatione intestinorum praecipue, tum et de amnio spurio, aliisque partibus embryonis gallinacei, nondum visis, observationes, in ovis incubatis instituta. *Novi Commentarii Academiae Scientiarum Imperialis Petropolitanae*, 12, 403–507.

Wolff, C.F. (1769). De formatione intestinorum praecipue, tum et de amnio spurio, aliisque partibus embryonis gallinacei, nondum visis, observationes, in ovis incubatis instituta. *Novi Commentarii Academiae Scientiarum Imperialis Petropolitanae*, 13, 478–530.

Zimmermann, J.G. (1774). *Traité de l'expérience en général et en particulier dans l'art de guérir.* Vincent, Paris.

Secondary sources

Canguilhem, G. (1955). *La formation du concept de réflexe aux XVIIe et XVIIIe siècles.* Presses Universitaires de France, Paris.

Carriou, J.-Y. (2019). *Histoire des démarches scientifiques. De l'Antiquité au monde contemporain.* Editions Matériologiques, Paris.

Cunningham, A. (2010). *The Anatomist Anatomis'd. An Experimental Discipline in Enlightenment Europe.* Ashgate, Farnham.

Daremberg, C. (1870). *Histoire des sciences médicales comprenant l'anatomie, la physiologie, la médecine, la chirurgie et les doctrines de pathologie générale.* J.-B. Baillière, Paris.

Darmon, P. (1977). *Le mythe de la procréation à l'âge baroque.* J.-J. Pauvert, Paris.

De Angelis, S. (2013). Experiments, judicial rhetoric and the testimonium. Practices of demonstration in the Hamberger-Haller controversy on the respiration mechanism. In *Scholars in Action. The Practice of Knowledge and the Figure of the Savant in the 18th Century*, Holenstein, A., Steinke, H. Stuber, M. (eds). Brill, Leiden, Boston.

Duchesneau, F. (2012). *La physiologie des Lumières. Empirisme, modèles et théories.* Classiques Garnier, Paris.

Entralgo, P.L. (1972). *Historia universal de la medicina. Volume 4: Medicina moderna.* Salvat Editores, Barcelona.

Foster, M. (1901). *Lectures on the History of Physiology during the Sixteenth, Seventeenth and Eighteenth Centuries.* Cambridge University Press, Cambridge.

Franklin, K.J. (1949). *A Short History of Physiology.* Staples Press, New York.

Grmek, M.D. (1990). *La première révolution biologique : réflexions sur la physiologie et la médecine du XVIIe siècle.* Payot, Paris.

Grmek, M.D. (2014). *Histoire de la pensée médicale en Occident. Tome 2 : de la Renaissance aux Lumières.* Le Seuil, Paris.

Guyénot, E. (1941). *Les sciences de la vie aux XVIIe et XVIIIe siècles.* Albin Michel, Paris.

Hall, T.S. (1969). *Ideas of Life and Matter. Studies in the History of General Physiology 600 B.C. – 1900 A.D.* University of Chicago Press, Chicago.

Pichot, A. (1993). *Histoire de la notion de vie.* Gallimard, Paris.

Pichot, A. (2011). *Expliquer la vie. De l'âme à la molécule.* Quæ, Versailles.

Ratcliff, M.J. (2009). *The Quest for the Invisible. Microscopy in the Enlightenment.* Ashgate, Farnham.

Richard, J. (1999). *L'expérience, l'observation et l'expérimentation dans la littérature médicale française du XVIIIe siècle.* ANRT, Lille.

Roger, J. (1971). *Les sciences de la vie dans la pensée française du XVIIIe siècle.* A. Colin, Paris.

Rothschuh, K.E. (1973). *History of Physiology.* Krieger, Huntington.

Salomon-Bayet, C. (1978). *L'institution de la science et l'expérience du vivant. Méthode et expérience à l'Académie royale des sciences 1666-1793.* Flammarion, Paris.

Sprengel, K. (1815). *Histoire de la médecine depuis son origine jusqu'au XIXe siècle*, volumes 3–5. Deterville, Paris.

Taton, R. (1995). *Histoire générale des sciences. Volume 2 : la science moderne.* Presses Universitaires de France, Paris.

Vignais, P. (2006). *Science expérimentale et connaissance du vivant. La méthode et les concepts.* EDP Sciences, Les Ulis.

West, J.B. (2012). *Essays on the History of Respiratory Physiology.* Springer, New York.

Wolfe, C. (2019). *La philosophie de la biologie avant la biologie : une histoire du vitalisme.* Classiques Garnier, Paris.

4

The Visible and the Invisible: Images of the Body, From Descartes to the Encyclopedists

Paolo QUINTILI
Università degli Studi di Roma Tor Vergata, Rome, Italy

4.1. The legacy of humanism and the Renaissance

The scientific revolution of the 16th and 17th centuries in Europe opened up two simultaneous paths for human knowledge in the field of life sciences, broadening its perspectives from both a methodological and a conceptual/theoretical perspective. Methodologically speaking, this expansion took place through the abandonment of a wise and hermetic conception of knowledge as a privilege reserved for the "select few" and the parallel adoption of an entirely experimental process – even though magical residues remain the legacy of the Renaissance era, with its animism and vitalistic naturalism – in the investigation of the world of life.

Conceptually, this expansion would be achieved through the development of new *models* of representation of living beings, particularly the organic body, which would attempt to make the image of the actual process of generation visible, as it occurs in nature. These rational models, even if produced through *hypothetical and conjectural* means, no longer have much to do with bodily qualities or qualitative

History of Life Sciences and Medicine 2,
coordinated by Céline CHERICI.

notions that are intangible and unimaginable to the average person[1]. In short, between the 16th and 17th centuries, human knowledge became "scientific" to the extent that it became *universal*, that is, open to everyone without discrimination.

As Paolo Rossi rightly observed:

> The three ideas at the heart of the new image of knowledge are: 1/to access science and truth, it is not at all necessary to despise that part of human nature which is "only human", nor is it necessary to belong to the selected few or the enlightened: it is enough to simply belong to the human species; 2/the processes for accessing truth, or the methods, are not complicated, but rather "modest", "simple" or "humble": these methods can be explained using very clear language and, for this reason, are – in principle at least, if not in fact – accessible to all; 3/all men can therefore access knowledge and truth. Knowledge is merely the explanation of possibilities that exist in all men. Science (as Arnauld and Nicole will say in *Éléments de géométrie*) consists only in "further carrying what we know naturally" (Œuvres, XLII, 215). (Rossi 2000, pp. 52–53 [our translation])

This clarification of the premises for the construction of a new knowledge of living beings, which will involve these new models and procedures for field research, is essential before examining this field, which is both ancient and revolutionary, in the 17th century: knowledge of the body and the organism, with different ways of representing them.

First, the *human* body takes precedence within the framework of *knowledge about living beings* in general: the latter conditions the representation of the former from a conceptual point of view and directs and shapes it (Canguilhem 1971, pp. 101–128).

The lesson of the Renaissance is of great importance here. The human microcosm is a small reflection of the macrocosm of the universe. It is from man and within man that knowledge of the infinite universe, and of the divine connected to it, originates.

1 The most eloquent example of this wise, hermetic and exoteric approach to knowledge of nature is the "chemical philosophy" and the work of Paracelsus (Philipp Aureolus Theophrast Bombast von Hohenheim, 1493–1541) and his disciples, which would have a significant impact in the 17th century; see Paracelsus (1983, 1955).

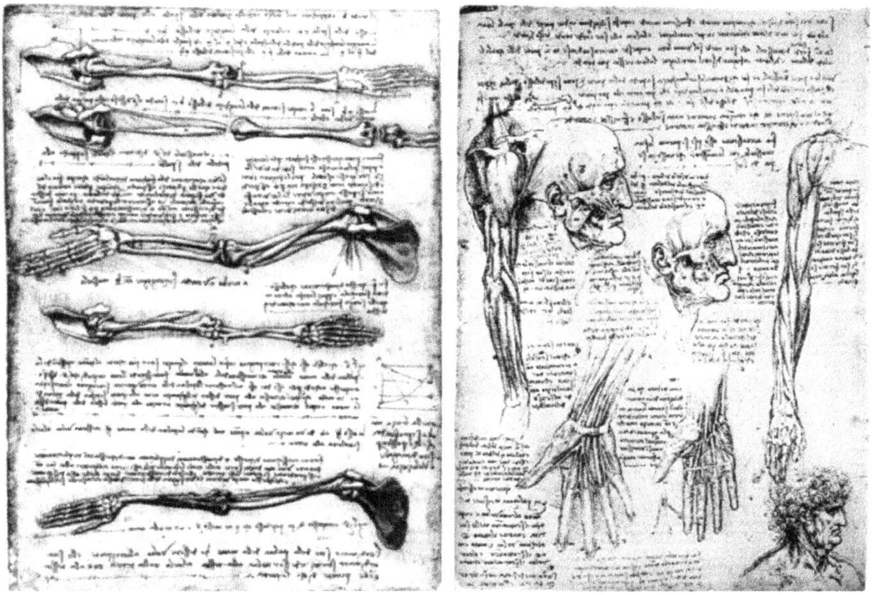

a) b)

Figure 4.1. *Leonardo da Vinci, drawings of the hand and arm with detailed descriptions, plate. 19000v/FA I v/c. 1510. Royal Library, Windsor (Le Nen and Laulan 2010, p. 32)*

The anatomical writings of Leonardo da Vinci (1452–1519) were the first to represent the historical and epistemological milestone that gave access to knowledge of the body, combining *art* – as the practical and operational mastery of the forces of nature for the production of artifacts – and *science*, as rational experimental expertise using observation to accompany this mastery. One example in particular is characteristic of this attitude: the anatomy of the human *hand*[2]. Leonardo focused his attention on this part of the human body to grasp its essence of spirituality and the intelligence of the universe as an equivalent of the soul of the world *in man*. It is "the anatomical organ that best expressed the rational nature of man", as Laurenza points out (ibid., p. 14), following the same Galenic tradition. Leonardo's drawings – "a mental operation, an eminently theoretical and rational act" – are still extremely precise and beautiful when compared to today's CT scans and the

2 "The default position of the palm of the hand is to face the horizon, and its usual extreme position is to face the sky or the ground, meaning toward the head or feet of the individual". (Le Nen and Laulan 2010, p. 33)

electronic "sketches" of the 21st century. We are witnessing the first operations of modern science, the inaugural gesture, one might say: the *decomposition* of the whole into its constituent parts, combined with direct first-person observation and reasoning, to better describe the latter and grasp their anatomical and physiological significance. Leonardo's *Codex Windsor* (da Vinci 1510) is undoubtedly the founding monument of modern anatomical science, preceding the work of Andreas Vesalius (Andries Wytinck van Wesel, 1514–1564) and his *Fabrica* of the human body (Vesalius 1543, 1st edition; 1555, 2nd edition). This Leonardian model prevailed among great anatomists, such as Vesalius and his successors in the 16th and 17th centuries (Cherici and Dupont 2019, pp. 55–84). The images in Vesalius' *Fabrica* are evidence of Leonardo's teachings, particularly with regard to the description of the bones and muscles of the hand and arm.

In his work, *De Omnibus Humani Corporis Interioribus Membris Anathomia* (de' Liuzzi 1316), which was published only in 1478 in Padua, the famous anatomist of the late Middle Ages – and Vesalius' predecessor – Mondino de' Liuzzi (1270–1326) presented images that were closer to theatrical figures or *diagrams* than to rational content. A famous example is the figure in the 1493 edition in Leipzig, where a doctor is depicted on his chair, watching while a barber operates on the body.

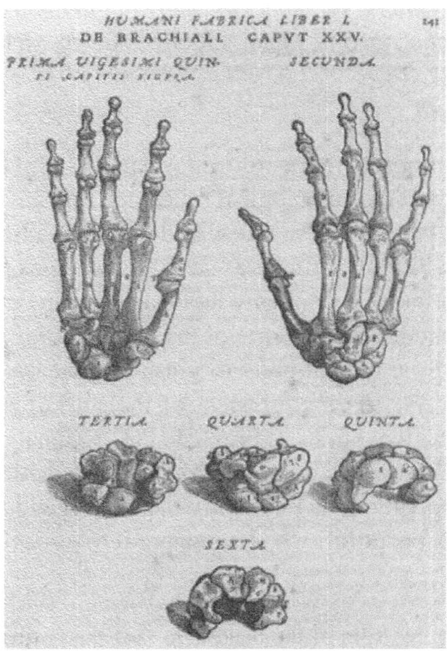

Figure 4.2A. *De Humani Corporis Fabrica (Vesalius 1543, p. 157, 219)*

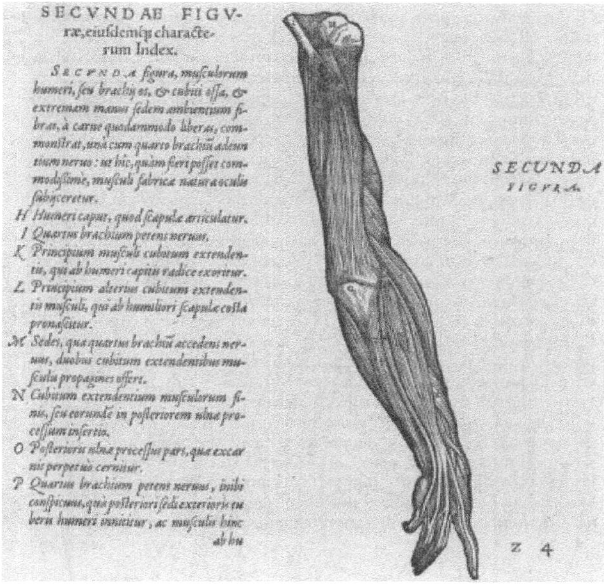

Figure 4.2B. *De Humani Corporis Fabrica (Vesalius 1543, p. 157, 219) (continued)*

Figure 4.3. *Mondino, De Omnibus Humani Corporis Interioribus Membris Anathomia*

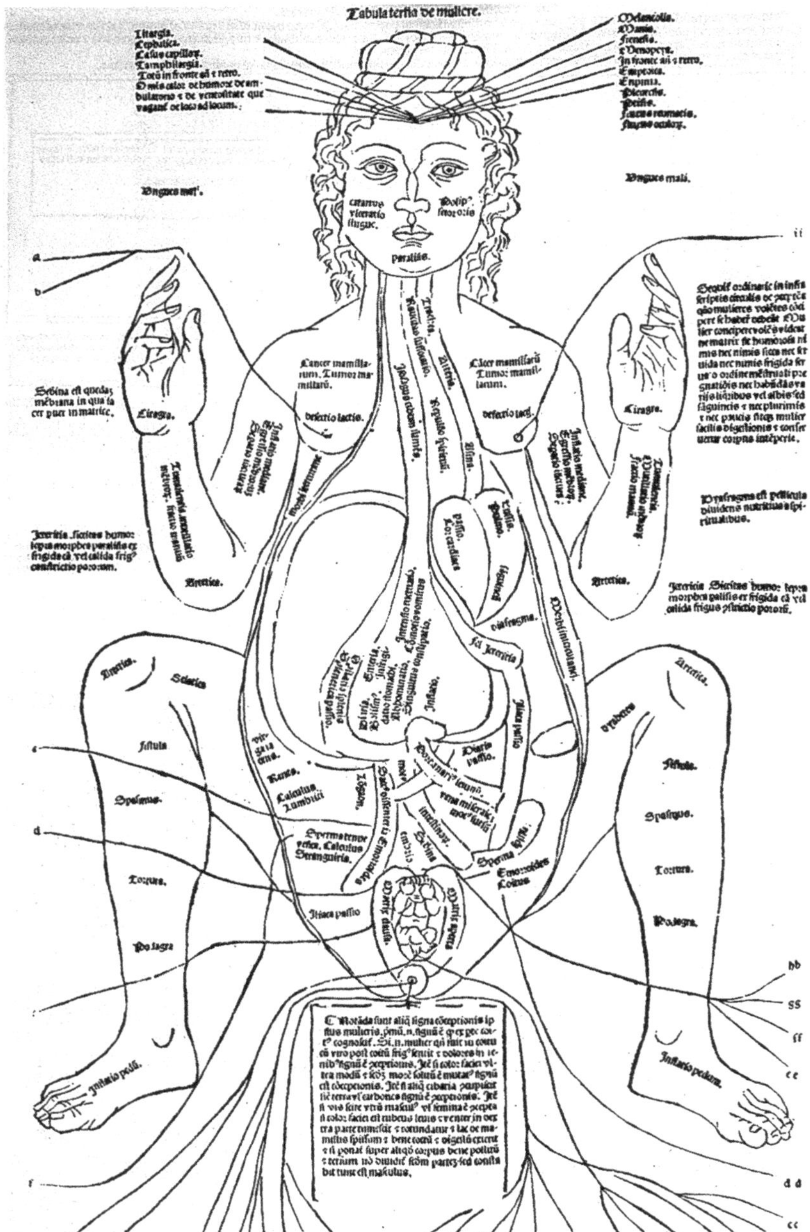

Figure 4.4. *Ketham, Fasciculus Medicinae, Venice, 1491 (Persaud et al. 2014, 57–58. o.)*

Even the images in *Fasciculus Medicinae* from the late 15th to the early 16th centuries, by the German physiologist Johannes de Ketham, published in Venice in 1491, provide illustrations that are purely logical *diagrams*. The image of man is the polar opposite of Leonardo's version, which is almost contemporary.

The difference between these two types of images – Leonardo/Vesalius and Mondino/Ketham – is obvious and more specifically consists of the union between art, science and experimental observation, which was realized in Leonardo's work and eventually became characteristic of the mentality of the scientific revolution, asserting itself definitively in the 17th century. The legacy of humanism and the Renaissance in the 17th and 18th centuries, in terms of the forms of representation of the body, therefore lies in the rejection of the approximations of hermetic imagination, the exoteric influence of various types of magical and astrological thinking on the basis of "correspondences"[3], and the acquisition/use of illustration techniques derived from the mathematical science of perspective and proportion. According to Leonardo and Vesalius, the *visible* aspect is now linked to a fascination of what can be measured and, in the case of the Baroque period, to the powerful and intangible attraction of that which is excessive, monstrous and extraordinary: ultimately, the *invisible*.

4.2. Separation of the microcosm and macrocosm: the Baroque period

Representations of the "zodiac man", which originated in the European and Arab Middle Ages, were widespread until the dawn of the modern era and inspired the representational schemes found in Mondino and Kethel until the 16th century. This is an intellectual legacy from an era in which the microcosm of humanity found its correspondences, resonances and "harmonies" in the divine macrocosm of the universe. Kepler (1571–1630) still adhered to this mentality, which one might call *holistic*, which permeated Vesalius' era. The representations and images of the body produced by the great anatomists and physician-philosophers of the Baroque period caused a break that led to a discontinuity between the microcosm and the

3 Such as the famous multiple representations of the "astrological man", an image in which each part of the human body corresponds to a sign of the zodiac (Persaud et al. 2014, p. 54): "Astrological Man. Wellcome Ms. 40. Calendarium 1463. (Wellcome Institute Library, London.) In the Middle Ages and during the Renaissance, astrology was taught in several European universities and many physicians were also astrologers. It was believed that the planets ruled different parts of the human body. Thus, Jupiter – lungs, liver and feet; Mars – left ear and genitalia; Sun – right side of heart; Venus – neck and abdomen; Mercury – arms, hands, shoulders and hips; Moon – left half of body and stomach". See: https://commons.wikimedia.org/wiki/File:Folding_Almanac_Wellcome_L0010970.jpg.

macrocosm. The breaking point was marked by the advent of *analytical* artistic depictions. The entire representation of the living body began to be shared, divided, sectioned and broken down into minute details and *superimposed layers*.

The Baroque era was a period of transition. As Deleuze observed, "the Baroque period does not refer to an essence, but rather to an operative function, to a trait. It never stops folding. It does not invent anything: there are the folds that came from the East, the Greek, Roman, Romanesque, Gothic and classical folds… However, it curves and recurves the folds, pushing them to infinity, fold upon fold, fold after fold" (Deleuze 1988, p. 5). The "folds" of the living matter represented curve and fit into one another, from the visible to the invisible. The image of the organic body, in the works of the natural philosophy of this period, attempts to make this dynamic structure intelligible. Since the Italian physicians of the late 16th and early 17th centuries, representation work has been faced with this new task. Not only to divide, section and anatomize but also to visibly show, in a manner of speaking, the invisible part contained in the observed reality.

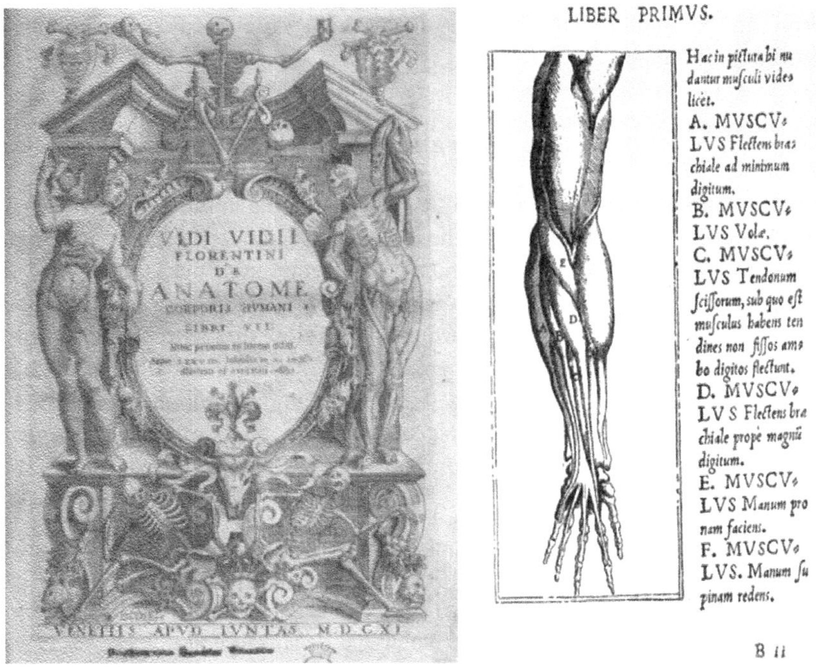

a) b)

Figure 4.5. *(a) Guidi, De Anatome corporis humani; (b) Canano, Musculorum humani corporis picturata dissectio*

We can cite the works of Guido Guidi (*Vidus Vidius*, 1509–1569), with his *De Anatome corporis humani* (Guidi 1611); Giambattista Canano (1515–1579), a contemporary and colleague of Vesalius, and author of a significant work on myology (Canano 1541); Matteo Realdo Colombo (1516–1559), author of *De Re Anatomica* (Colombo 1559); Costanzo Varolio (1543–1575), with his research on optic nerves (Varolio 1573), followed by a posthumous *Anatomia* (Varolio 1591); Giulio Cesare Casseri (1552–1616), a great anatomist from Piacenza, a specialist in the organ of hearing (Casseri 1600–1601) and the author of a posthumous work on anatomy (Casseri 1627), which soon became famous, and reminiscent of the work of the Paduan master *Hieronymus Fabricius*, a student of Gabriele Falloppio (1523–1562); Girolamo Fabrici d'Acquapendente (1533–1619), the most influential anatomist of the late Renaissance, with his numerous studies in comparative anatomy (d'Acquapendente 1603, 1623, 1909).

a) b)

Figure 4.6. *Casseri, Tabulae anatomicae: larynx, heart and tongue; formation of the fetus*

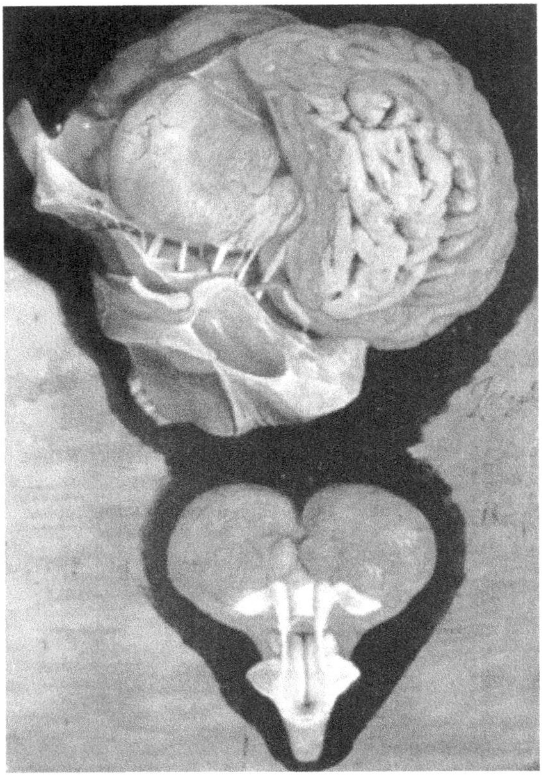

Figure 4.7. *D'Acquapendente, Tabulae Pictae, the lateral sulcus. For a color version of this figure, see www.iste.co.uk/cherici/history2.zip*

Delving into the microscopic folds of living body parts is always a long, slow and deep *descent*. Anatomists in the early 17th century taught René Descartes the very technique of the "mechanism" of the human body through their methods of representing and dissecting the living body. Perhaps the most remarkable example of this Baroque trend (Carlino 2010; Carlino and Ciardi 2012) of delving into the miniscule infinity of bodily folds is the work of Pietro Berrettini da Cortona (1596–1669), who was not a doctor but a Roman artist, painter, architect, and author of a posthumous work on anatomy (da Cortona 1618), published only in 1741 in Rome. In Pietro da Cortona's plates, the structure of body image, with its multiple semantic levels of meaning, is clearly visible through the division of work (Figure 4.8).

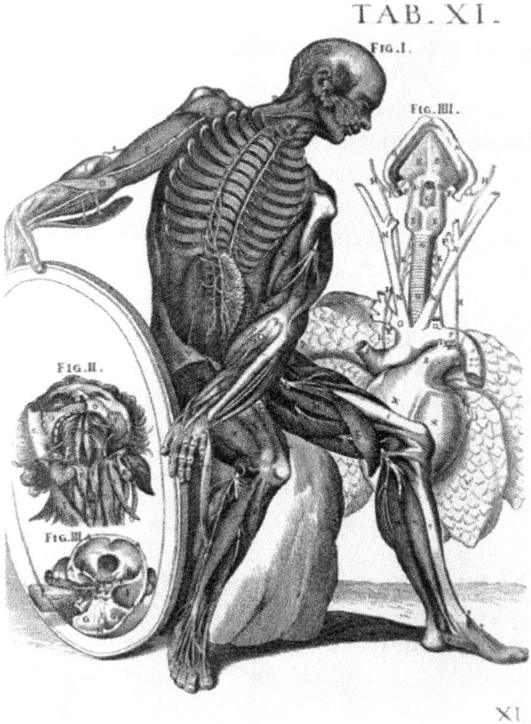

Figure 4.8. *Pietro da Cortona, Tabulae anatomicae, XI*

Figure 4.8 illustrates a skeleton showing its full interior, including muscles, tendons and skin; then, "descending", in the sense of a magnifying mirror, toward the detailed depictions of the trachea and respiratory system; the figure then describes, as a whole, the interior of the smaller parts of the same system (heart, trachea, lungs, etc.), with anatomical and physiological rationalization, marking the relationships between these parts (indicated by letters) and their functions. We do not know the anatomist with whom Berrettini collaborated to produce these images in 1618, which were published only a century later for the first time; however, he was undoubtedly a Roman contemporary, demonstrating a taste for dramatizing a dissection that aims to astonish the reader with the dazzling spectacle of the body and death. The man dissected in front of the broken mirror of himself was not yet understood in the body/mind duality that would define him in Descartes' time. He is an enigma, or rather a dream of himself, waiting to be revealed. For this reason, this style of representation, which uses the aggressive perspective of fragmented bodies to express Baroque concerns about the mystery of life and death, has been called

"Dream Anatomy". Through this style, physicality is captured in terms of dissolution, folding and withdrawal into itself, according to Deleuze's interpretation.

Pietro Berrettini da Cortona thus accomplished an essential step in the history of anatomy, toward the emancipation of what Renaissance humanists called the human "microcosm", which now plays a separate, special and individual role in relation to the "macrocosm" of the physical universe. The body becomes a new macrocosm. There is an emphasis on the spatialization and individuation of the body and its parts, which plunge into the depths of the infinitely small, so to speak, and thus find technical autonomy in relation to the soul/mind that inhabits it, decisively transcending the religious, anthropological, psychological and sociological taboos (related to the fear of desecrating the dead) that had previously been a focus in anatomical practice (Quintili 2019, pp. 13–15). With Descartes, this transition to the autonomy of the body was definitively accomplished: as a pure machine composed of countless springs and gears, such as *res extensa*, physical matter would provide new forms of representation in the mid-17th century.

4.3. The Cartesian body machine: between Harvey and Descartes

William Harvey (1578–1657), who had been a student of Fabrici d'Acquapendente in Padua, the "discoverer" of blood circulation (Harvey 1628), directly inspired the physiological and biological work of René Descartes (1596–1650).

However, Harvey was not the only one to produce significant medical, anatomical and biological works that marked this transition toward technical autonomy in the field of body representation, linked to the spread of Cartesianism. This approach – also called the "Cartesian mentality" since the second half of the 17th century – of seeing the organic body of humans and non-humans as an autonomous physical reality ontologically worthy of knowledge, from which the various moral and theological concerns of past centuries, has been removed (Carlino 1994, 1998; Quintili 2019). Andrea Carlino rightly emphasized the presence of a "metaphorical imperative" of anatomy here, from the Middle Ages to the modern era, which weighed heavily on this science and which had to "redeem" itself for the sacrilege of having desecrated the body through the act of dissection, by using guilt-absolving metaphors inscribed in images (Carline 1999). The Cartesian turning point, which rejected this heritage, was, in this sense, a revolution for the emerging biological sciences. The body (*res extensa*) and soul (*res cogitans*) are two distinct and unique substances, each with its own characteristics. Each substance, in its own domain, obeys specific, different laws: those of thought for the soul and those of matter and movement (heat), namely, physical mechanics, for the body. No mixing or confusion of the two fields is possible or valid in true science.

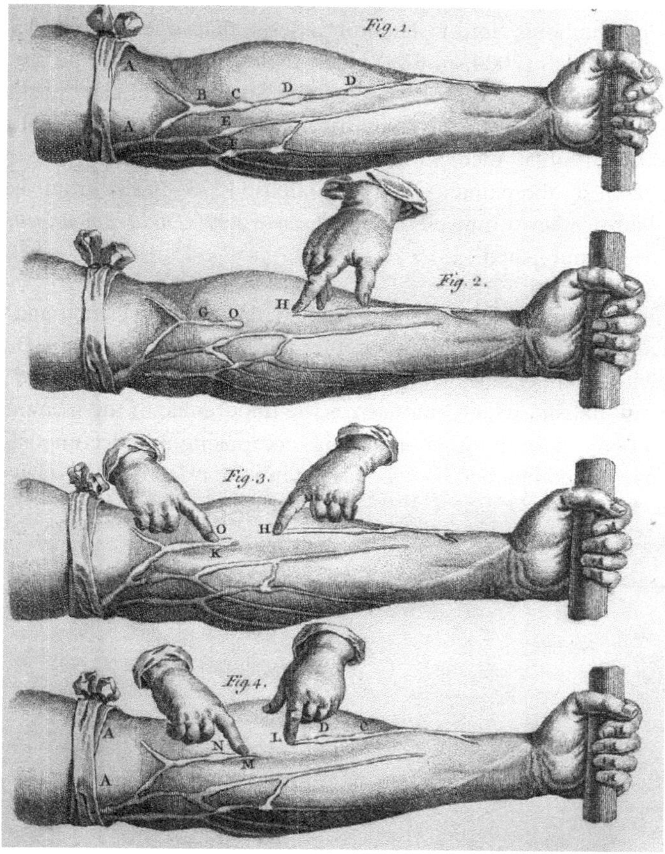

Figure 4.9. *Harvey, De motu cordis (Harvey 1628)*

This revolution, known as the "Cartesian revolution" – although it was not accomplished by Descartes alone – enabled the technical division of the image of the body to flourish and the "geometric" method to be applied to the specialized study of its parts, according to the laws of matter and physical mechanics (matter and movement).

In the Dutch communities of the Great Century, which Descartes had elected as the place for his research, there was a leading anatomist who mastered the art of dissection with great creativity and whom he probably knew: Adriaan van den

Spiegel (1578–1625), known in the Latin tongue as Spigelius, was a student and successor of Giulio Casserio at the chair of anatomy in Padua, the European capital of anatomy at the time, where he also died. In his *Éléments de physiologie* (Diderot 1773–1784, posthumous; ibid.), Diderot asserts that one should "always have Spigelius at your side" (2004, p. 389) when listing the books to be studied for his work. *De Humani Corporis Fabrica* by van den Spiegel, published posthumously in Venice in 1627, is perhaps the most significant collection of anatomical plates, after Vesalius, being of great medical and artistic value because of the magnificent engravings drawn by the painter Odoardo Fialetti (1573–1638), from the Tintoretto workshop which Casserio himself intended for a new *Theatrum anatomicum*, and which remained unpublished.

The combination of Italian (Venetian) art with Flemish science and technology produced this unique work, blending together two worlds: Berrettini's Baroque style (associated with Mannerism), with its taste for macabre staging, and the age of mechanistic rationalism, which emphasizes the importance of the meaningful details of the human body, describing, enumerating, composing and decomposing it before the reader's eyes, accompanied by scholarly symbolic references.

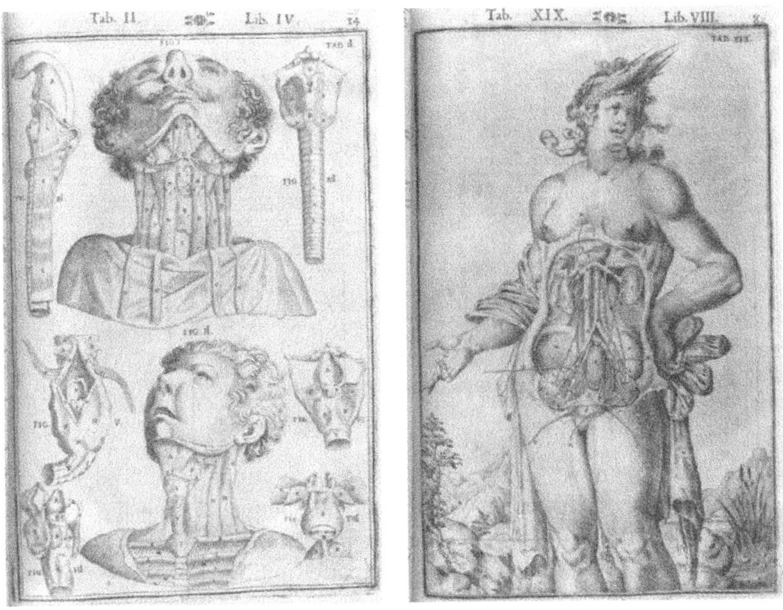

a) b)

Figure 4.10. *Spigelius, De Humani Corporis Fabrica, the trachea; female reproductive organs (the goddess Ceres)*

The plates in another work by Spigelius, *De Formato Foetu*, on pregnancy and fetal anatomy, published in Padua in 1626, then posthumously in 1631, deepen and broaden this perspective, from the whole picture to the tiniest parts, which, through Fialetti's eyes, takes on a Venetian style, with a kind of "flower child" aesthetic that makes the parts blossom before the reader's eyes (van den Spiegel 1626 [2nd edition 1631]; idem 1627 [2nd edition 1632]).

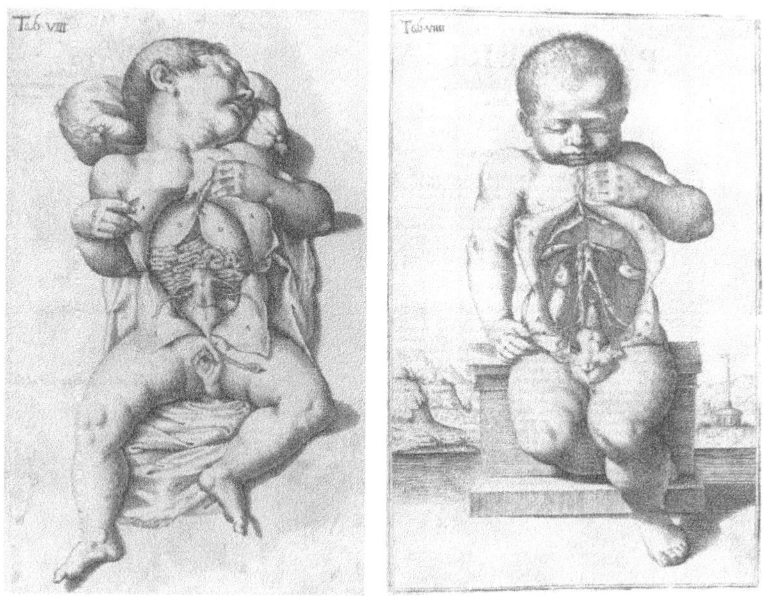

a) b)

Figure 4.11. *Spigelius, De formato fœtus, tables VIII–IX*

The Baroque taste for presentation is still present, but there is also no concern for respecting the metaphorical imperative concerning the body, even that of a child subjected to the profane act of dissection. It is a body similar to any other body dissected in front of a previously unseen *theatrum anatomicum*. Spigelius' lesson had a long-lasting effect, and it is quite likely that Descartes himself – in addition to Spinoza, who had it in his library – was familiar with this model of anatomical representation. The impact that this new mechanistic anatomy had, or may have had, on Descartes' thinking is difficult to assess, especially since the images of man and the body appearing in Cartesian works are not of great iconographic importance. In the two posthumous editions of René Descartes' *L'Homme*, published in Latin in Leiden in 1662 and in French in Paris in 1664, the figures presented were invented.

Nevertheless, the opposite remains true: in the subtlest way, and in terms of *philosophical theory*, Descartes' thinking recorded what had seemingly been achieved for decades – by Fabrici d'Acquapendente, Casserio, and Spigelius – in the practice of anatomical science, with its process of autonomization of the technical image that we have looked at in the previous paragraphs. The image of the body, as much as the science on which it depends, thus made significant progress in the first half of the 17th century and began to free itself from its ties to theology, morality and politics.

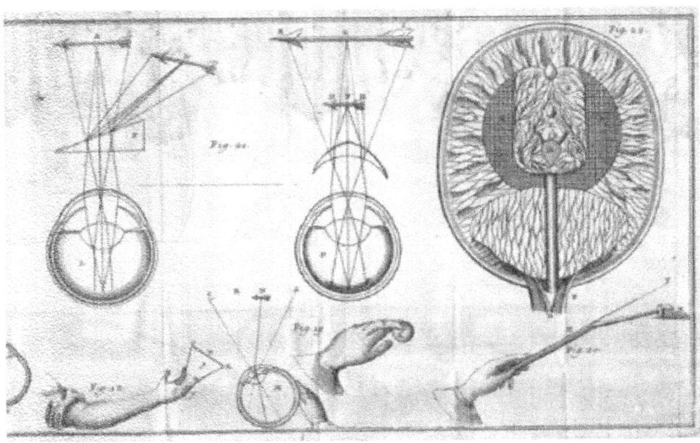

a)

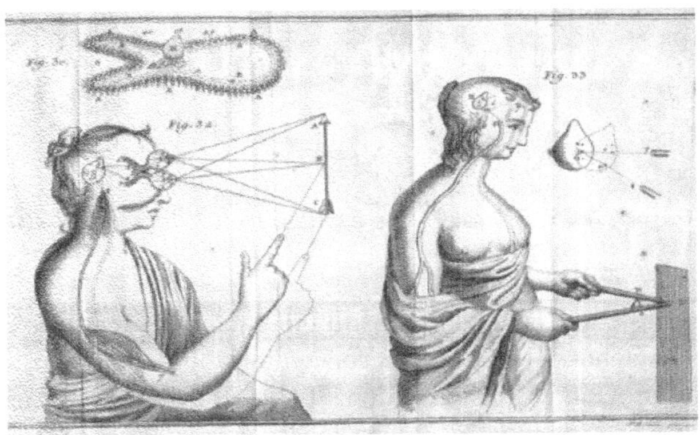

b)

Figure 4.12. *Descartes' L'Homme (Descartes 1996), the eye, vision, the brain and the pineal gland*

The author of the illustrations in the French edition of *L'Homme* was Gerard van Gutschoven (1615–1668), a great Flemish anatomist and mathematician whom the publisher, Clerselier had met in Louvain. He was part of the generation of students of Spigelius and Casserio and was very comfortable with Descartes' new mechanical philosophy. Louis de La Forge (1632–1666), a Protestant theologian from Saumur and a doctor of medicine, living in the town of La Flèche – Descartes' birthplace – had also been consulted and was consequently also involved in the project. The two artists thus worked together on the illustrations for this *Treatise* (the letter G indicates the hand of van Guschoven, and the letter F that of La Forge, who worked on fewer images), marking a threshold in the history of Western culture, more for its philosophical content than for the nature of its illustrations, which follow the recent tradition of anatomical engraving. Man, as a physical being, is a pure "machine made of earth" (Chapter 1: "*De la Machine de son corps*"):

> Men will be composed, like us, of a Soul and a Body; and I must describe to you, first the body separately, then the soul separately [...] I suppose that the Body is nothing more than a statue or machine made of earth... (Descartes 1996, p. 1)

Descartes adds that his objective is not to describe organs that are already well known to anatomical scholars but rather to explain their movements, actions and relationships. He therefore provides a mechanical explanation of the functions that result solely from the simple interaction of organs, comparing it to the interaction of the parts of an automaton. In his conclusion, Descartes emphasized this meaning. The body cannot be removed from the "Great Mechanism" that is the law of the world. The human body is no longer a microcosm but rather a natural being, among others, composed of microcosms or "cogs", without any difference in quality or substance: *res extensa*, governed by the laws of mechanics, matter and motion. This conquest was also achieved thanks to the anatomical works that appeared at the beginning of this century, which was rightly called "Cartesian". Foucault correctly speaks of a "Cartesian moment" (Foucault 2001, p. 15).

4.4. The successors of Descartes and the Flemish anatomists

There are still traces of the "metaphorical imperative" of anatomy, even after the completion of the Cartesian revolution, which, like all revolutions, always leaves behind (and ahead) survivors, remnants of bygone eras. In this context of transition, a special place is occupied by the work of the anatomist Aimé Bourdon (1636–1706), the *Nouvelles Tables anatomiques* (Bourdon 1678), who worked in Cambrai, protected by his patient, the archbishop of the city, Jacques-Théodore de Bryas (Bourdon 1678). During his lifetime, he had connections with the School of Montpellier. This work involves a new way of representing the body, following Descartes, *revealing the soul* that is inseparably linked to it. This is a representation that already belongs to what could be called the "spirit of Enlightenment of the 18th century", a spirit that emphasizes the unity of the "two substances" by representing a physical presence – *a wind* – in the body, which passes through the system of veins, nerves and the brain, emanating from his figure, from right to left.

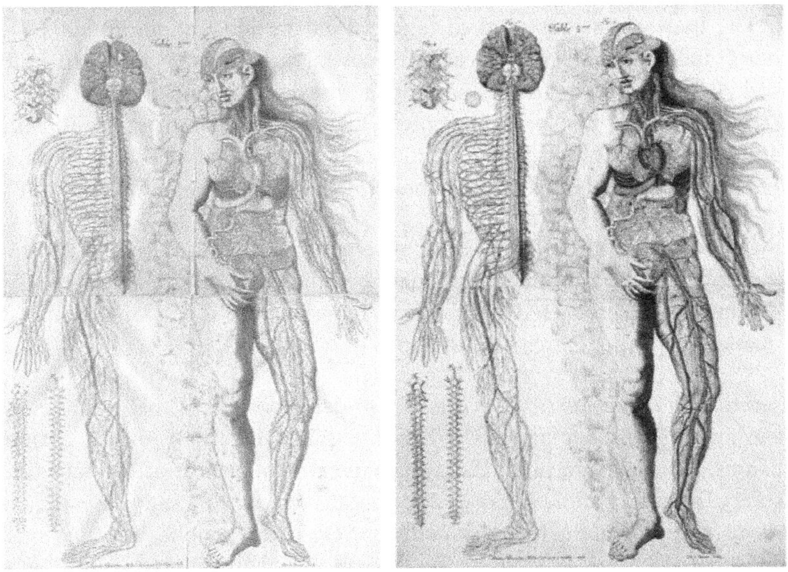

a) b)

Figure 4.13. *Bourdon, Nouvelles Tables anatomiques, table VIII (two versions). For a color version of this figure, see www.iste.co.uk/cherici/history2.zip*

Bourdon focuses on the details of the image to suggest delving deeper into the soul within this same physical figure. He provides several versions, several facets. The anatomist's intention was to show the different ways in which water escapes from the body. However, the duality of the figure, which leads to unity of the soul and the body, is symbolically represented by the light *breath of wind* that moves from the network of nerves toward the dissected figure, the naked, crying girl and toward her organs, which she hides with her hands. This female figure then blossoms in the almost immaterial waves of her hair, from the soul to the body. Bourdon's woman cries because she has probably been condemned to death, and her physical form must remind us of this origin, which redeems her once she has become an object of science. This is a new, previously unseen version of the metaphorical imperative of anatomy, expressed here at the highest level and freed from its regard for religion.

Until Bourdon's time, anatomists and illustrators, in a way, had to purge the guilt associated with the major conquest of medical science: the dissection of bodies (intended for the Resurrection, but now devoted to science) that could only be legally carried out on the corpses of society's outcasts – those condemned to death, suicides, adulterers, etc. The dissected and decomposed body is therefore always linked to the need to restore the idea of wholeness and "redemption", which places the objective on a symbolic and metaphorical level (the myth of Marsyas, for the act of dissection, Adam and Eve, etc.). The *beauty* of wholeness must therefore always contrast the horror of bodies torn into pieces, in the flayed corpses that were associated with the guilt-absolving reference to punishment and suffering.

It was therefore *essential* to reconstruct, through images, something we could always recognize as a human body in its entirety, imbued with moral meaning through metaphor. It is as if the dissected and anatomized human being were taking symbolic revenge through an iconography that does not merely represent skeletons and flayed bodies as if they were inert objects, exposed like pieces of meat on a butcher's block.

During the Age of Enlightenment, this metaphorization, which was still obviously going through artistic representation, such as in the Renaissance, was nevertheless less and less marked by religious restrictions. It became less and less concerned with reconstructing wholeness. The body machine and the man machine were finally in the spotlight, freeing anatomy from its sense of guilt, de-metaphorizing and, in a way, de-symbolizing the parts of the body that were broken down into their smallest components. It was no longer necessary to restore the wholeness of the human body. The physical man can ultimately be seen as a "system of organs" (de Bordeu 1751; Diderot 2004, pp. 334–347), governed by the founding principle of the *sensibility* and irritability of living matter (Albrecht von Haller).

A Flemish anatomist, who was a contemporary of Bourdon, also marked this transition to a Cartesian representation of the body, playing around with the metaphorical imperative to break free from it, in a manner that would later become typical of the Age of Enlightenment: Govert Bidloo (1649–1713). His work, *Anatomia Humani Corporis*, was published in Latin in Amsterdam in 1685 and was republished in Danish in 1690 (Bidloo 1690), accompanied by plates engraved by the Flemish painter Gérard de Lairesse (1641–1711). The latter was the last representative of classicism at the turn of the "Golden Age" and an old friend of Rembrandt, who painted his portrait. As Raphaël Cuir has rightly noted, Bidloo and Lairesse's representations mark "the death of the flayed body or Cartesian anatomy" and the birth of a completely unfinalized medicine and anatomy[4]. In Bidloo's work, the anatomized body is no longer a flayed body that bends and unbends more or less elegantly in a Baroque style but rather a *corpse* opened under the attentive eye of the doctor, who describes and lists its parts.

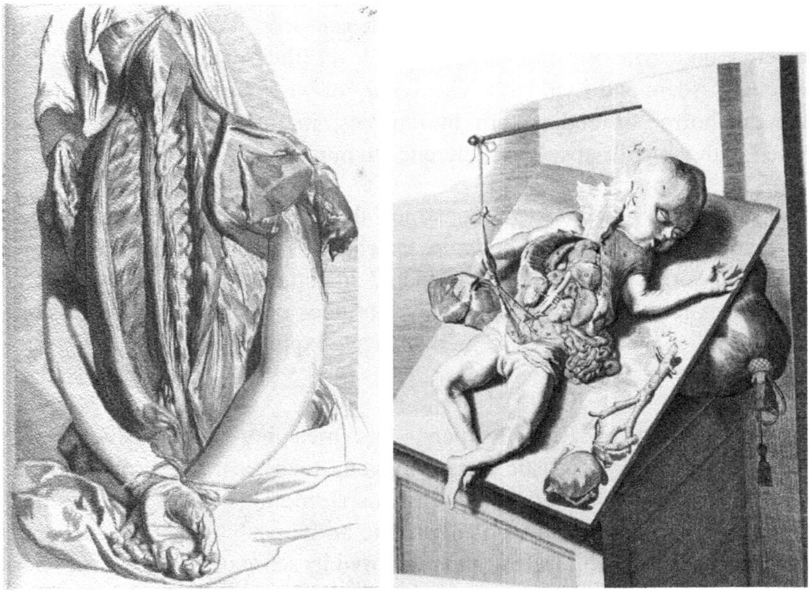

a) b)

Figure 4.14. *Bidloo, Ontleding des menschelyken lichaams, tables 30 and 59*

In the 105 plates of their work, *Anatomia*, Bidloo and Lairesse, who worked side by side, gave little space to the external setting of the image, focusing with great

4 Cuir (2019, p. 41): "With Descartes and the development of Cartesianism, the body closes in on itself, while beginning to multiply. For the anatomist, the body is no longer the body experienced in everyday life, the body is no longer the body for the artist. The microcosm and the macrocosm no longer mirror one another, and the dissected body gradually leaves the picture to inhabit the 'objective' abstraction of the blank page. God can even leave the body once the machine is complete, his work is more about the soul. By rejecting theology, by separating the soul from the body and by assimilating the latter to a machine, renouncing the analogy between the microcosm and the macrocosm, Descartes isolates the body from all environments, from all symbolic connotation, to make it an object that is strictly devoted to study. In other words, with Descartes, the body is disenchanted".

intent on the smallest details and, above all, on the dissector's technique, by very tightly framing the corpse's body. In the two plates above, we can see the juxtaposition of an image that is still laden with ambiguity: that of a bound and dissected woman, a not-so-distant reminder of the metaphorical imperative that required the details of the ropes on the wrists to be shown, symbolizing the punishment that the condemned person had to endure. However, the representation of the body is completely free from the taboo of dissection. In fact, next to it, the dissected fetus of a 7-month-old girl shows her internal organs without any concern for metaphorical retribution for this figure of innocence, which is almost martyred by the anatomist's action, devoted to the higher purposes of scientific knowledge.

It should also be remembered that Bidloo's work draws heavily on the epistemological model provided by his master, another remarkable Flemish anatomist, Frederik Ruysch (1638–1731), who has been aptly described as the "artist of death" (Kooijmans 2011). A large canvas by Jan van Neck (1634–1714), entitled *The Anatomy Lesson of Doctor Frederick Ruysch* (1683), shows Ruysch in the center of the image dissecting the corpse of a child. Like the more famous *Anatomy Lesson of Dr. Nicolaes Tulp* (Rembrandt 1632), this painting dramatically reproduces the same act but in a more analytical and mechanical manner, without characters, of the anatomist Bidloo, his student, who is depicted dissecting this fetal corpse in table 59 of his *Anatomia*.

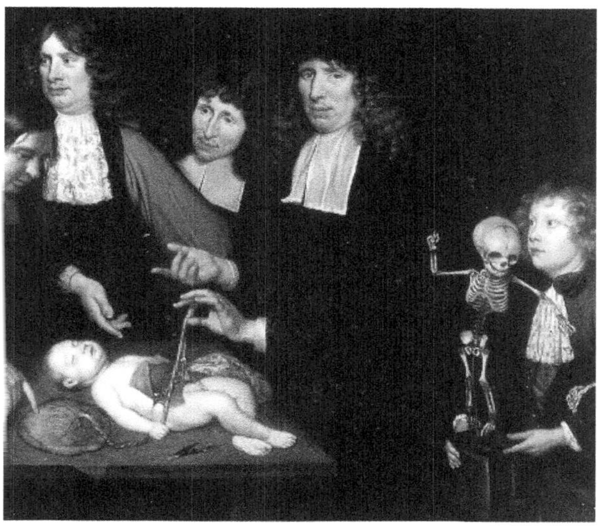

Figure 4.15. *Van Neck, The Anatomy Lesson (details). For a color version of this figure, see www.iste.co.uk/cherici/history2.zip*

There is a clear semantic echo between these two images. Ruysch and his anatomy are perhaps the best examples of this "Cartesian moment" (Foucault) in medical science in the late 17th century. Ruysch's work is distinguished by its focus on the body at two prominent moments in its life: birth and death. Ruysch created numerous "mummies" using his new embalming and organ preservation technique, which enabled him to create a genuine natural history museum in his private home in Amsterdam. These are "picturesque" anatomical preparations consisting of isolated and preserved organs, as well as fetuses (ibid., Chapters 4–6). The resulting image of the body is a reproduction of the same dynamic of dissolution and the division of its parts, captured over time, as if frozen by the art of anatomy. Here, anatomy no longer uses the arts but rather becomes an art in its own right. The art of freezing death, of stopping death in images, thus makes it "resurface" (Hansen 1996, pp. 663–679).

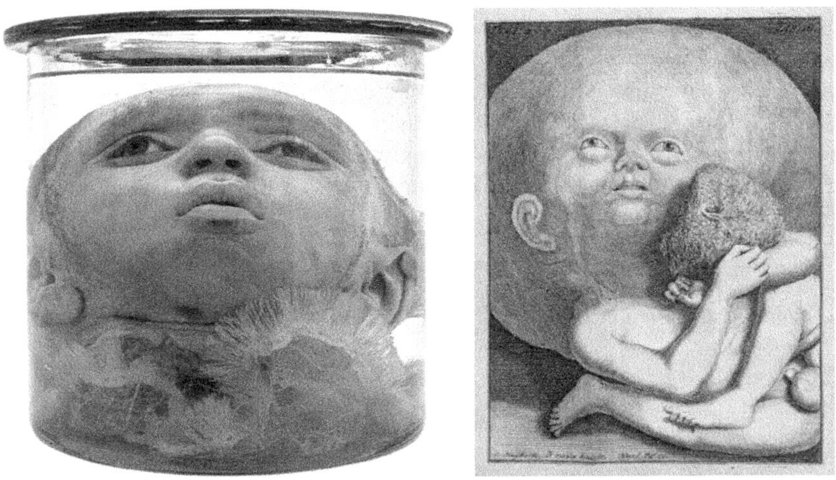

a) b)

Figure 4.16. *Ruysch, Thesaurus anatomicus I, plate III, child's head with injections. For a color version of this figure, see www.iste.co.uk/cherici/history2.zip*

Ruysch's "baroque" taste for the macabre has been mentioned (Guest 2012), but the rather "Cartesian" aspect of the representation in the *Thesaurus anatomicum* consists of the construction of veritable "anatomical machines", composed of living "cogs", sectioned, divided from the whole body, and recomposed by the physician/artist (Ruysch 1702–1715, 1716, 1721). A large part of Ruysch's collections and compositions were purchased in 1716 by Tsar Peter the Great and are now kept in the museum in Saint Petersburg[5]. The Tsar was fascinated by this art (and power) of mastering death through the new technique of injections – using very subtle liquid wax – and the preservation of specimens of limbs and organs in decorated glass jars – using alcohol-based liquid solutions – which froze images in the final state of the dead body, as if it were still alive[6]. Thus, Ruysch constructed real anatomical theaters through images, with “mummies” under glass and compositions of skeletons. He was particularly interested in the bodies of fetuses and stillborn babies. The author of the drawings and engravings is the Danish painter Cornelis Huyberts (1669–1712), a very creative artist.

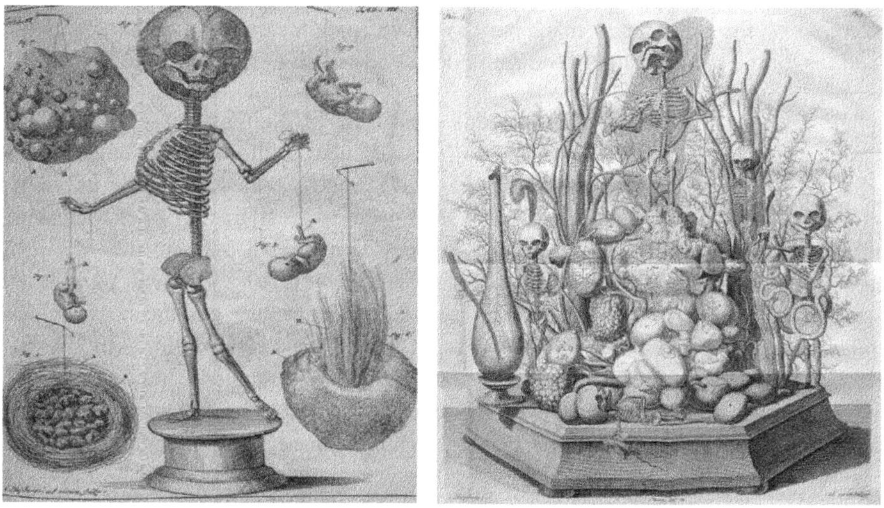

a) b)

Figure 4.17. *Ruysch, Thesaurus anatomicus VI, table III; Thesaurus anatomicus I*

These representations were of major interest for two reasons. On the one hand, to develop his technique, Ruysch focused on the microscopic aspects of body parts and tissues, which he devoted a large part of his *Thesaurus* to; the technique of injections – which he had learned from his friend, the microscopist Jan Swammerdam

5 See: www.kunstkamera.ru/en/museum/kunst_hist/5/5_2.

6 Ruysch’s work also inspired the poet Giacomo Leopardi (1798–1837), who dedicated one of his *Operette Morali* (Leopardi 1827) to him, entitled *Dialogo di Federico Ruysch e delle sue mummie*; see Leopardi (1992).

(1637–1680) – enabled him to see (and depict in his tables) what had never been seen before: the smallest vessels and muscle fibers, the mesentery and other organs separated in their tissues, etc. The injection was a research tool that would be of great significance in the future (Golgi), even for anthropometric preservation techniques. On the other hand, Ruysch focused his attention on the parallels between animal and human anatomy. His magnum opus, the *Thesaurus*, in the posthumous edition of 1727–1744, begins with a *Thesaurus animalium* in a single volume, followed by the *Thesaurus anatomicum* in ten volumes. Ruysch also depicts human limbs or organs alongside animal organs or placed in the same preservation jar, showing real consideration for comparative anatomy, subject to the same Cartesian criteria: the dissection of the body into "machinery" and the parallel explanation of forms and functions, or the preservation of the entire dead body to give it a kind of new life[7].

The first to inherit and perfect this approach to anatomy was the Flemish physician Steven Blankaart (1650–1704), a student of Bidloo in Amsterdam, who published *Anatomia reformata* (Blankaart 1687), a work that was already published in Danish in 1678 and was dedicated to Bidloo himself. Blankaart openly declares his anatomical work to be Cartesian, with full awareness, from the *Praefatio* discourse of *Anatomia*, addressed to the reader. This is a work program, the "reform" of the discipline that the author wants to propose, on the basis of two fundamental principles: (1) the body is a "clock" or an android composed of detachable and separable parts: *rotas* ("cogs") whose *synthesin* ("composition") must be known; and (2) the circulation of blood is the main phenomenon from which this "composition", or multilateral relationship, can be grasped and understood

7 Fontenelle's testimony: "The parts were injected in such a way that the last branches of the vessels, finer than a spider's thread, became visible, and, what is even more astonishing, sometimes only with a microscope; what substance could be so fluid as to penetrate such channels, while simultaneously being so solid as to harden there? Small parts could be seen that are not visible in either the living or the freshly dead. [...] Everything that was injected retained its consistency, softness, and flexibility, and even became more beautiful with time, because the color became more vivid to a certain extent. [...] Everything was protected from corruption by Mr. Ruysch's secret. A very long life gave him the pleasure of seeing none of these pieces spoil with age, and of being unable to determine their lifespan. All these dead bodies, with no apparent drying, no wrinkles, a rosy complexion, and flexible limbs, were almost resurrected; they only appeared to be asleep, ready to speak when they awoke. Mr. Ruysch's mummies prolonged life in a way, whereas those of ancient Egypt only prolonged death" (Fontenelle 1731, pp. 102–103). This praise undoubtedly inspired Leopardi (see below).

clearly and distinctly[8]. The plates clearly show this assembly trait, the "texture", the *synthesin* that is the "clock" of the human body.

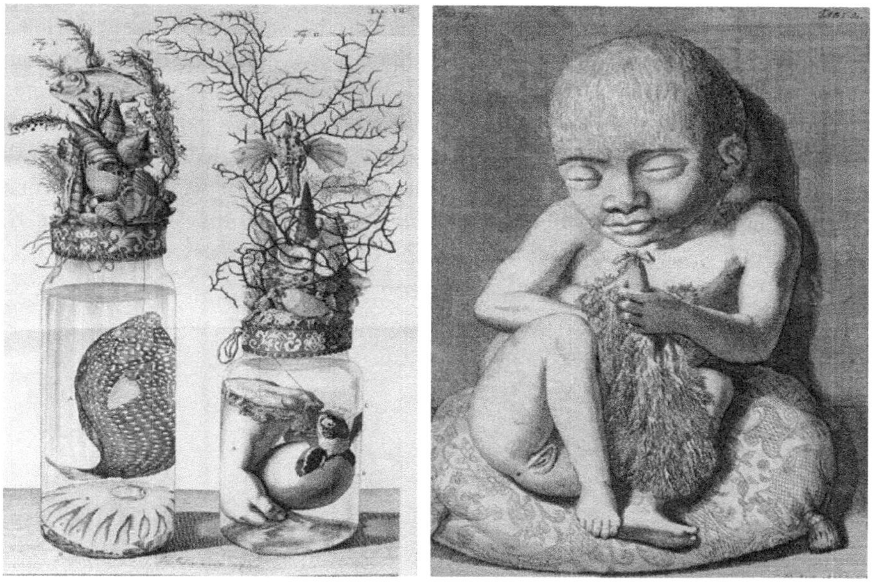

a) b)

Figure 4.18. *Ruysch, Theatrum animalium, plate VII; Thesaurus anatomicum III, plate I (fœtus)*[9]

8 "Caput I. *De Anatomia, ejusque subjecto ac divisione.* Nemo artis Medicae intima recte perscrutari potest, nisi corporis Humani exquisitam habeat cognitionem. Quaemadmodum enim imperi-tus alicujius Horologii, ipsius συνθεσιν ac compaginationem ignorat, etiam cum deficiat, resti-tuendo non erit : ita nullus, rectam morborum notitiam habebit, ignota corporis nostri machina, nemo novit horologii impedimenta, nisi qui rotas, fusos distinguere, ac quomodo ab elatere seu primo mobili motus concitentur, optime soiat. Adeo oportet eum qui Medicinam facere vult, *subjectum* suum, seu corporis humani texturam, rite rimari" (Blankaart 1687, p. 1).

9 Ruysch adds a poem in Latin to the description of plate I: "Quid sumus, aut quidnam nostri post funera restat? Adspice : nil aliud, nuda nisi ossa, vides calculus, et vastum lapidum, quem cernis, acervus, quam si vita hominum plena dolore, docet. Exhibet hoc tabulis, hoc verbis explicat, urbis Ruischius Amstellae laus, medicumque decus" (Ruysch 1721, p. 5), which also represents a "research program": to determine, through anatomy and science, "what remains" of the human being *post funera*.

The image that would later become the "machine man" a few decades later (La Mettrie 1747) is well illustrated in the work of Blankaart, who, incidentally, had already translated Descartes into his own language. This work by the Flemish anatomist can be linked to work carried out at the same time by a doctor in Montpellier on the strange and complex "separate body" that is the human brain and nervous system. This doctor was Raymond Vieussens (1635–1715). He wrote a *Neurographia Universalis* (Vieussens 1684), a work offering a parallel image of the "neuronal" man of the 17th century (Vieussens 1716), a detailed description of the invisible network of the nervous system, which is similar to the *second body* of man, an image that clearly appears in the double *in folio* plates of his *Neurographia*. His text was republished several times until the end of the 18th century.

a) b) c)

Figure 4.19. *Blankaart, Anatomia, plates LIX, LII, XXXIX: the digestive system and the throat*

In Vieussens' work, the method of breaking down the *machina* of the neuronal body and reassembling its parts is clearly Cartesian, a method openly acknowledged by the author. He discusses this in his *Epistle to the Reader* and asserts his originality for having applied the mechanical method to the description of the brain, nerves, and spinal cord, "described in vivo, illustrated by engravings, with their action and use explained by a physical discourse", as the title states. The distinctive feature of Vieussens' work, which follows in the footsteps of Ruysch and Blankaart, is that it made the invisible (the nerves and the tissues they penetrate) visible through the parallel use of dissection and the microscope. This instrument began to be used in anatomy and medicine at the end of the 17th century, when Marcello Malpighi (1628–1689), the true founder of human "microanatomy" (or microscopic anatomy), owing to the invention of the first microscope by Antoni van Leeuwenhoek (1632–1723) (van Leeuwenhoek 2014), made detailed observations of separate anatomical parts of the body, particularly the lungs. In his work *De pulmonibus* (Malpighi 1661), he discovered the true capillary structure of organs, which are composed of numerous "glands", the pulmonary alveoli. Malpighi also played an important role – aside from his fundamental research in embryology – for his work in the field of neuroscience, a few decades before Vieussens, with his work on the structure of the brain, *De Cerebro* (Malpighi 1665), in which he observed a close structural and biological homology between nerve matter and the brain (Adelmann 1966; Quintili 2008; Engelhardt 2021).

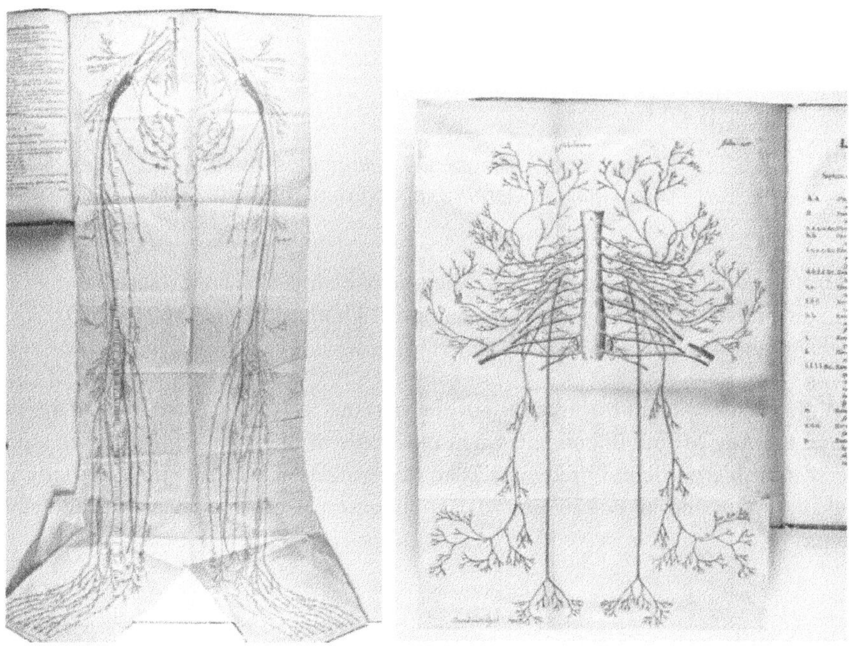

a) b)

Figure 4.20. *Vieussens, Neurographia Universalis (Vieussens 1716), plates XIX and XXIV*

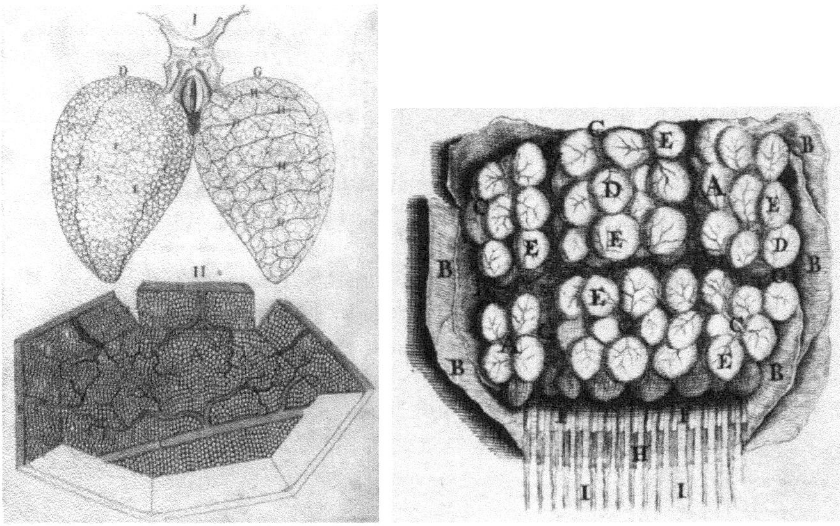

a) b)

Figure 4.21. *Malpighi, De pulmonibus, table II; Malpighi, De Cerebro, structure of the cortex*

On this point, concerning nerves, Vieussens claimed to have "seen" things that Vuillisio himself, the great English anatomist Thomas Willis (1621–1675), in his *Cerebri anatome* (Willis 1664), had failed to see[10].

With Vieussens and his anatomy of the nervous system, we are thus at the very edge of the Age of Enlightenment, when the image of man as a *natural* being among others, with his particular "machine" and "organization", in his unity of body and mind (the brain and its *compaginatio*), would assert itself in a conscious and stable manner.

10 (Vieussens 1716, Chapter "Lectori", n.p.): "Et quoniam nervos plures ad cutem, quam versus musculos, aliasque partes deferri observaveram, neque de ipsis quicquam ab auctoribus, *imo nec ab ipso Vuillisio*, qui hanc materiam periitus exhausisse videbatur, dictum fuisse jamdudum adver-teram, ea subiit animum cogitatio, ut Neurologiae partem illam, si fieri posset, excolerem, quam a caeteris Anatomicis fere intacta fuisse cognoscebam" (our italics); on the delicate question of "seeing" in anatomy, medicine, and art, which often led to error, see Snyder (2015); on the delicate question of generation, see Maienschein (2014).

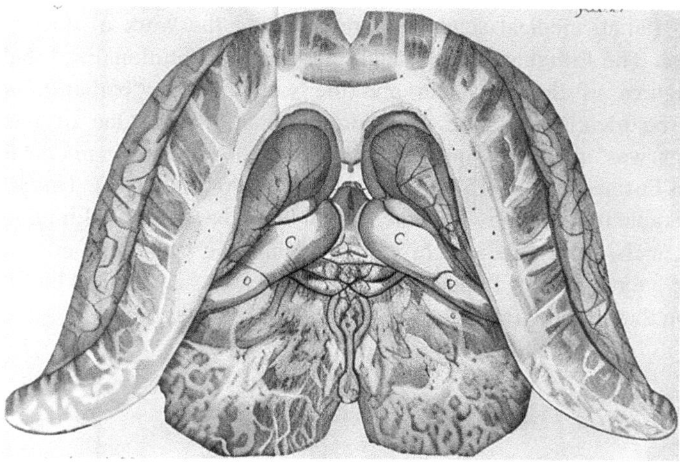

a)

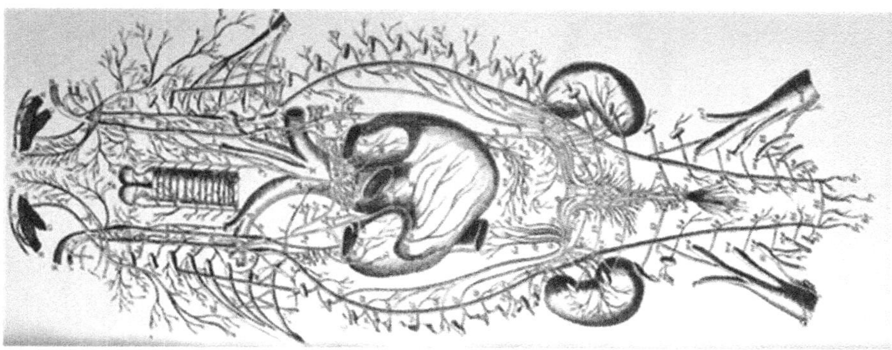

b)

Figure 4.22. *Vieussens, Neurographia Universalis, plates VIII and collapsible table outside the text*

4.5. Images of the body during the Age of Enlightenment: from Lancisi to the Encyclopedia

Giovanni Maria Lancisi (1654–1720) was perhaps the first physician to inaugurate the new medical and anatomical science of the Enlightenment in Europe. As a great anatomist, "practical" physician and founder of one of the world's richest libraries of medical texts – which is still open to the public today – Lancisi also founded an authoritative school at the University of Rome La Sapienza, which was recognized as being on par with that of Padua. The school worked to rebuild a living tradition of Italian medical science by re-assessing the work of doctors from the classical age. The Enlightenment brought about and institutionalized the joining of the two figures of the physician and the surgeon (the "barber"), who would henceforth be identified in the same professional figure. One of Lancisi's last undertakings was in fact the publication of the *Tabulae Anatomicae* in 1714 by Bartolomeo Eustachi (1500–1574) (Eustachi 1714; Fortuna 2014). This is one of the most remarkable representations of the human body that the early Enlightenment has given us. Lancisi updated Eustachi's plates, which he had found and purchased – through the intercession of Pope Clement XI, of whom he was the chief physician – in an edition that became famous and was reprinted several times.

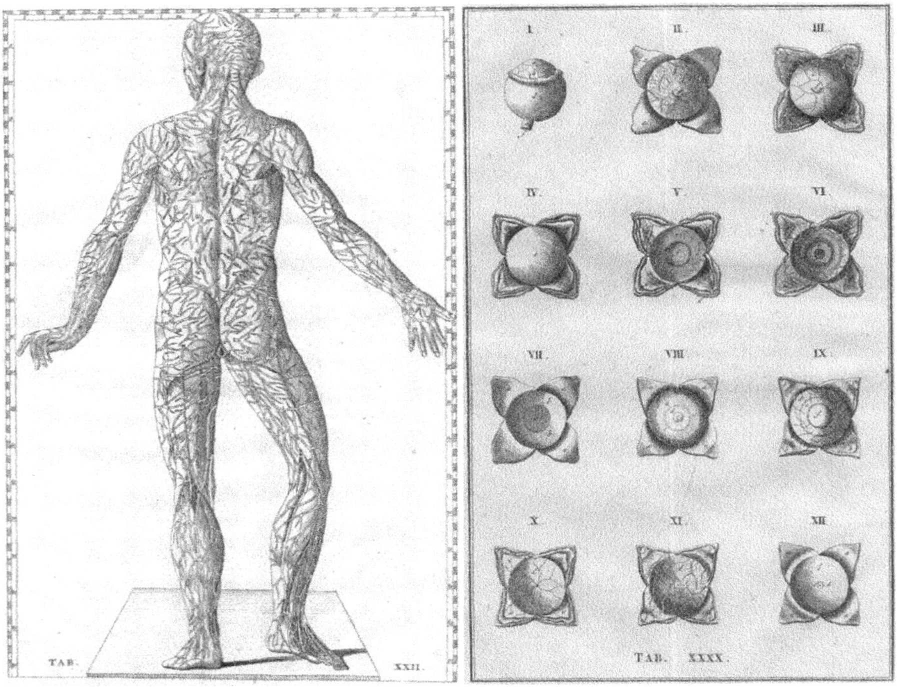

a) b)

Figure 4.23. *Eustachi and Lancisi, Tabulae Anatomicae, XXII (the vein system) and XXXX (the eye)*

De motu cordis, et aneurismatibus, Opus postumum in duas partes divisum, published 8 years after his death (in 1728), is Lancisi's anatomical masterpiece, especially in terms of the scientific significance of the discoveries made, such as the

"Lancisi sign", an abnormality in the circulation of the jugular vein. Adopting the iatromechanical method of the recent tradition of Malpighi and Antonio Maria Valsava (1666–1723), Lancisi proposed a true mechanical "composition" of the anatomical parts of the heart, starting from the description of the organ in a healthy state and ending with its placement in the cardiovascular system. Lancisi subsequently identified the main physiopathologies affecting circulation, particularly aneurysms.

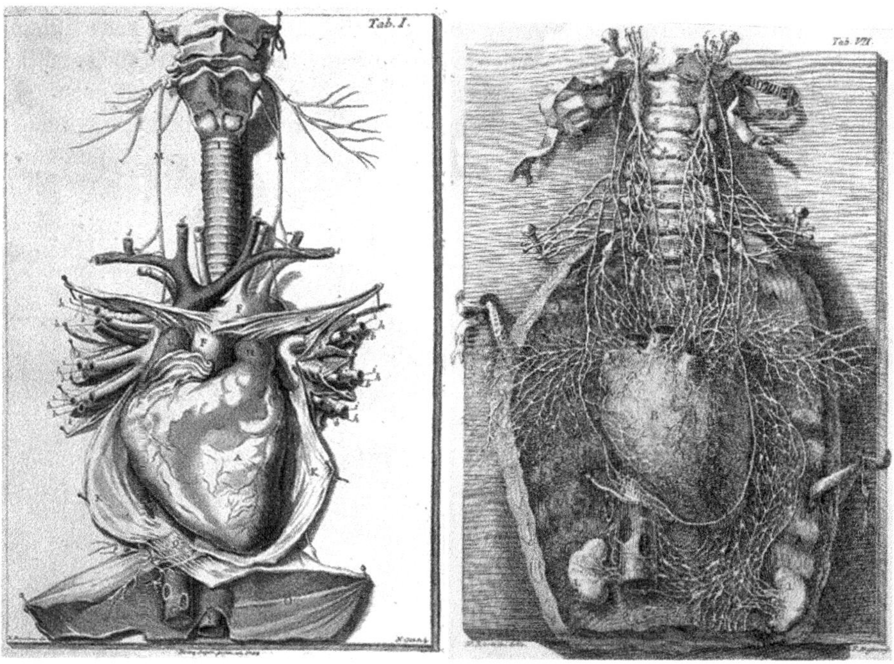

a) b)

Figure 4.24. *Lancisi, De motu cordis, tables I and VII*

Just a few years after Lancisi, other innovations ushered in a new anatomical representation of man in the Age of Enlightenment, again stemming from the Italian iatromechanical school of Malpighi and Valsava. Giovanni Battista Morgagni (1682–1771), a contemporary of Montesquieu and Diderot who lived between Bologna and Padua during the Encyclopedia era, made his mark on the history of the discipline with his work *Adversaria anatomica* (Morgagni 1706, 1762). This text was republished several times. Morgagni is the leading representative of what might be called the "medical enlightenment", with his critical work *reforming the anatomical gaze*, which is forced to activate its critical judgment in the face of what the eyes – and the microscope, in particular – offer in terms of observation. The new alliance between the eye and *critical reasoning* produced what is perhaps the finest example of an anatomical representation of the human body in the modern era. The *Adversaria* are, so to speak, the decisive victory of rational and experimental iatromechanical medicine in the Age of Enlightenment. Until Descartes and the first Flemish Cartesians, disease had still been considered a *holistic* phenomenon, following the old teachings of Hippocrates and Galen. It was an imbalance of bodily humors, a disturbance of the *harmony* between the different components of the body. With Morgagni – and a little later in France with Théophile de Bordeu and the Montpellier encyclopedists – disease became a specific matter for this relatively autonomous animal machine, with its own characteristics, called an "organ". Morgagni provided effective physiological descriptions of the organ, highlighting the various pathologies that could affect it.

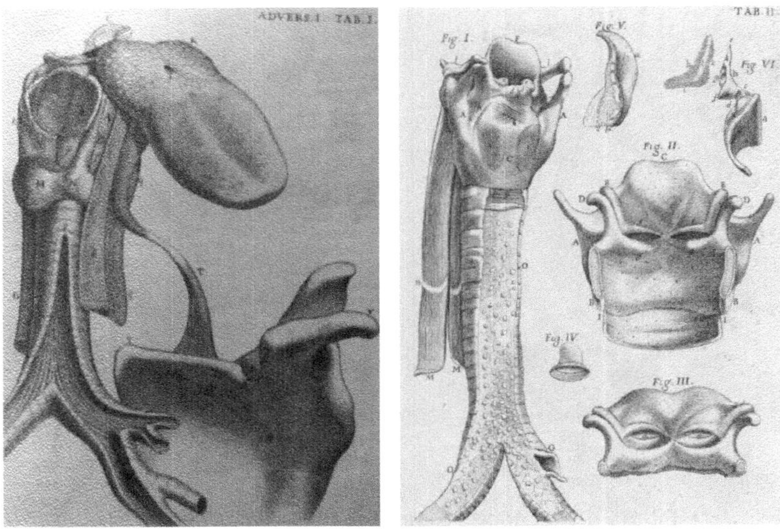

a) b)

Figure 4.25. *Morgagni,* Adversaria anatomica prima, *tables I and II (the language system and the glottis)*

In Morgagni's 1761 masterpiece, *De sedibus et causis morborum*, which is the founding text of modern pathological anatomy, it is the parts/cogs of the bodily machine that fall ill, not the person or the whole animal (Cosmacini 1997; Grmeck 2014; Tubbs et al. 2019). Morgagni demonstrates that each symptom can be shown, through the anatomy of the organs, as emerging from some specific pathological process underway in some particular, singular structure of the body. The true effectiveness of the "machine model" can be verified in the field of pathological anatomy. Consequently, the metaphor of the machine itself takes on a new meaning. This is no longer the old *hydraulic* machine of the Cartesian *Man*; here, we are dealing with the complex *weaving machine* (like the stocking loom), made up of tiny, invisible components intertwined in the organic texture, which the anatomist's eye must expose and which medical critical reasoning is expected to judge correctly. It is specifically based on this level of sound judgment that the revolution in pathological anatomy took place, marking the Enlightenment. Morgagni's approach would impose a new style in the representation of the body in images, which would be adopted by anatomists at the end of the century, such as Paolo Mascagni (1725–1815), who was Napoleon Bonaparte's physician and author of *Vasorum lymphaticorum corporis humani historia et iconographia* (Mascagni 1787), which can be closely compared to the last illustrations in Morgagni's *Adversaria Anatomica*.

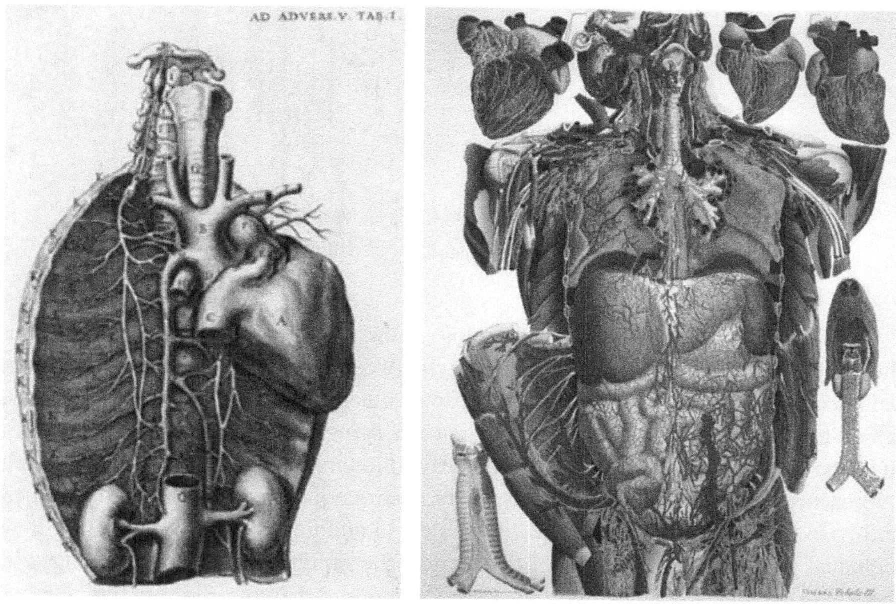

a) b)

Figure 4.26. *Morgagni, Adversaria anatomica V, I; Mascagni, Vasorum lymphaticorum, III. For a color version of this figure, see www.iste.co.uk/cherici/history2.zip*

The anatomical image – and the body that is depicted – became both more analytical and, simultaneously, more realistic on the basis of a combination of precise observation (of changes in the state of an organ) and critical judgment in diagnosis. Indeed, 17th-century microscopy often produced errors concerning the deformation of objects due to the polarization of light and the resulting distortion effects (Solinas 1967; Croft 2006).

In fact, many erroneous judgments had been made about the structure of spermatozoa, which were interpreted differently, in a whimsical manner – according to dominant preformationist theories (as depicted in Figure 4.27, the animal is contained entirely, as a miniature, in seminal matter) – by naturalists at the end of the 17th century, as can be seen in van Leeuwenhoek (1679), Hartsoeker (1694) and de Plantade (1699).

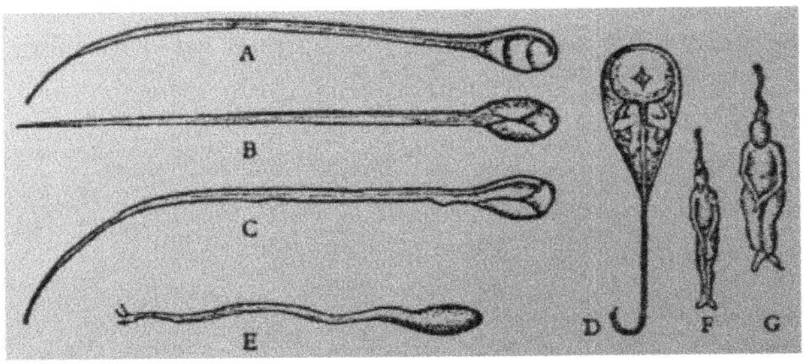

Figure 4.27. *Spermatozoids. A, B, C: van Leeuwenhoek; D: Hartsoeker; E, F, G: Plantade*

Van Leeuwenhoek (1722) was the most astute observer, even when interpreting the structural and morphological homology between the anatomical parts of human beings (sperm and fetuses), as well as those of other living beings, including plants, which belonged to the same order as *organized beings*. In anatomy, therefore, from the first half of the 18th century onward, following the example of the Dutch microscopist, it was necessary to have very accurate judgment and a highly refined ability to observe and interpret the "invisible", in addition to the availability of new technical means (Maienschein 2014), to grasp the minimal pathological changes in the body associated with morbid manifestations affecting the organs. Before Morgagni (1706–1716, 1761), all of these elements could not be brought together.

In fact, works from the turn of the 17th century clearly show the hybrid nature of these anatomical representations, which bridge an old world that no longer exists (the Renaissance and the *Vitruvian Man*) and another world that is struggling to assert itself under the weight of 1,000-year-old tradition. In England, this was the case with the English physician and politician James Drake (1667–1707), author of *Anthropologia nova* (Drake 1707), a work cited by encyclopedists related to *Anatomy* plates on arteries. Drake introduced a neologism that would be widely used in the future: "anthropology", limiting its meaning to the study of the physical man. His tables are remarkable and still express a vision that is, in a way, on the fence: finalist or holistic, and simultaneously analytical; however, the image of the human being is the product of a compromise between Vesalius and Descartes.

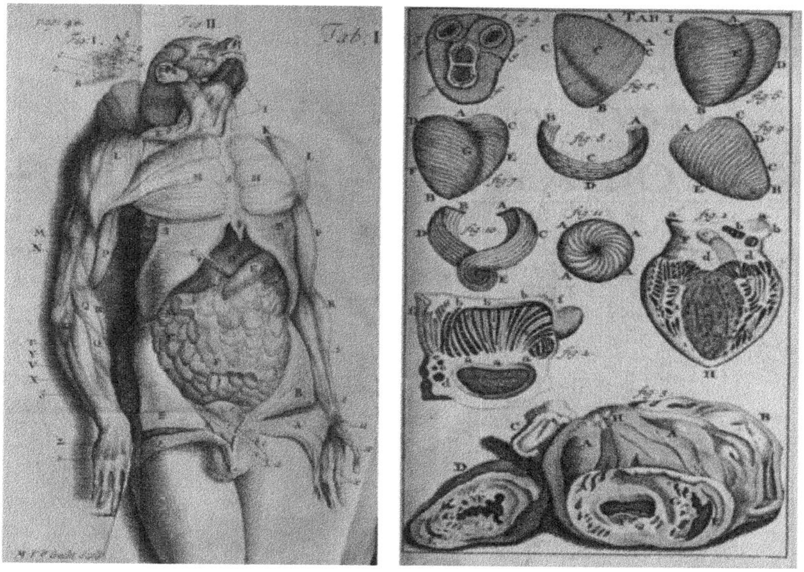

a) b)

Figure 4.28. *Drake, Anthropologia Nova, tables 1–2*

The same can be said of his contemporary English surgeon William Chelseden (1688–1752), who wrote two famous works on anatomy (Chelseden 1713) and osteography (Chelseden 1733[11]), and, in France, of the great heir to Ambroise Paré (1510–1590), Pierre Dionis (1643–1718), who was a surgeon to Louis XIV and

11 Chelseden was a friend of Locke and would give substance to the famous "Molineux question [or philosophical hypothesis]" (Locke 1690, II, 9, 8) concerning a blind man who regained his sight thanks to cataract surgery (which Chelseden performed successfully), and who had to recognize and name a cube and a sphere using his sight alone, without touching them. Chelseden's experiment was the inspiration for Diderot's *Lettre sur les aveugles* (Diderot 1749).

author of *Anatomie de l'Homme* (Dionis 1690, 1707), in which he considered "the latest discoveries" concerning the heart (Harvey, de Senac) and the nervous system (Willis, Vieussens, etc.). The value of Dionis' work lies in the fact that he also depicted the *female* body alongside the male body, the dominant model in the imagination of Western medical history.

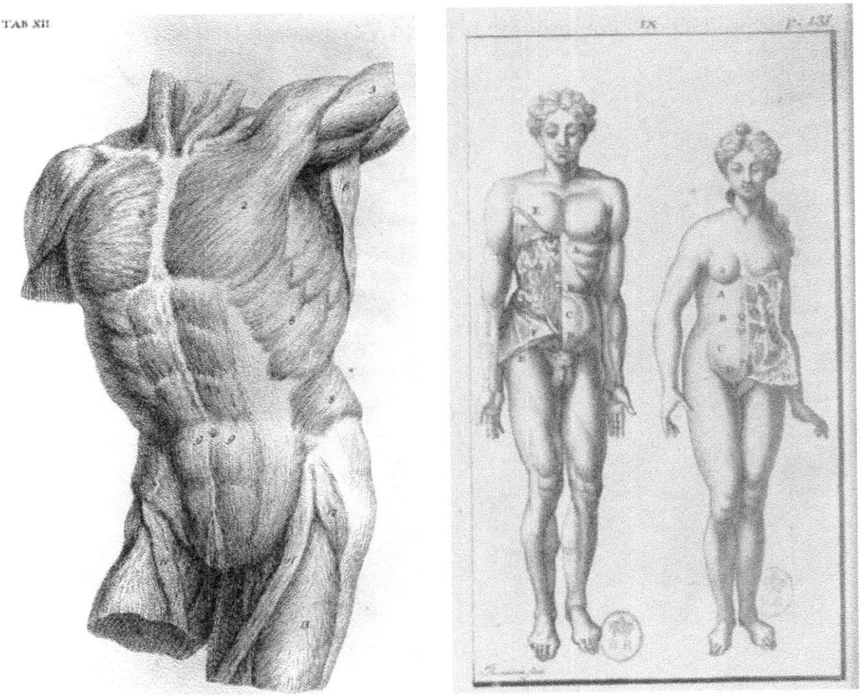

a) b)

Figure 4.29. *Chelseden, Anatomy of the Human Body, XII; Dionis, L'Anatomie de l'Homme, IX*

Diderot and d'Alembert's *Encyclopédie*, which was published in 1751 and continued to be published until 1772 (*Supplement*, 4 volumes, 1777–1782), produced a kind of overview, or compendium of the various anatomies seen up to that point, from Eustachius to von Haller. The same process had been adopted by Diderot for the *Description des Arts mécaniques*, whose images had been taken from other earlier technical descriptions. Diderot entrusted the publication of the 33 plates to Pierre Tarin (1725–1761), a young doctor from the École de Paris, who got on well with his colleagues and collaborators from Montpellier: Théophile de Bordeu

(1722–1776), Jean-Joseph Ménuret de Chambaud (1739–1815), Henri Fouquet (1727–1806) and Paul-Joseph Barthez (1734–1806). In addition to the images by Drake, Eustachi, Vieussens and others mentioned above, Tarin offers readers the works of German physicians who were less significant from the anatomical illustration perspective, with the exception of the great work of Bernhard Siegfried Albinus (1697–1770) (Albinus 1747; Elkins 1986). This anatomist, born in Frankfurt, was a student of Bidloo and Boerhaave in Leiden, followed by Ruysch in anatomy in Paris. Albinus' tables have two sides: one analytical and formal and the other artistic. The images are, in fact, on two *parallel* levels: the first shows the entire figure, artistically engraved; the second analyzes the different parts, piece by piece, indicated by dots.

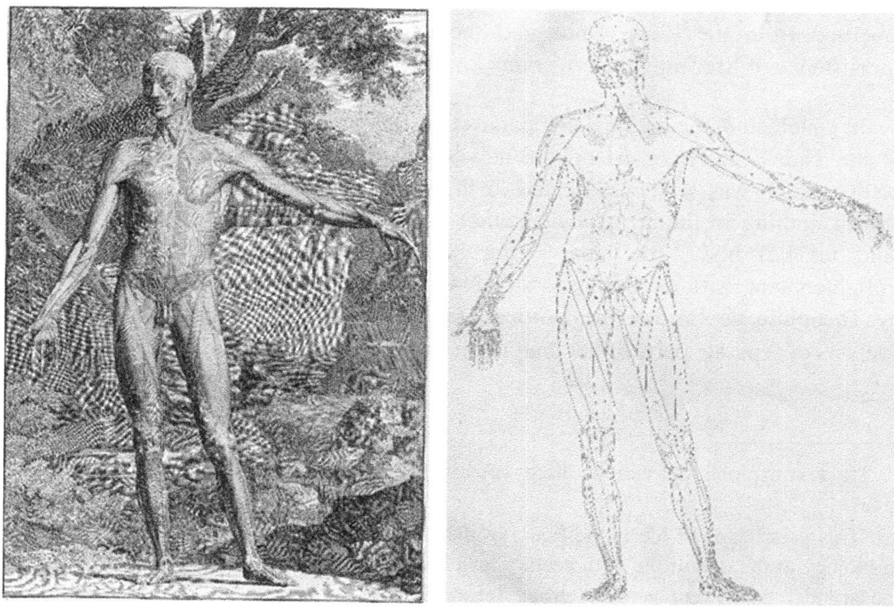

a) b)

Figure 4.30. *Albinus, Tabulae sceleti et musculorum, tables I-II (reproduced in the* Encyclopédie*)*

Diderot's *Encyclopédie* only includes the engraved "artistic" image here, leaving the explanation of the parts (which is quite difficult to follow) to the captions accompanying the plates. Other Flemish and German anatomists, some less well known or forgotten, were used by Tarin. First, Regnerus de Graaf (1641–1673) (de Graaf 1668), then Antonius Nuck van Leiden (1650–1692) (van Leiden 1690), Johann-Adam Kulmus (1689–1745) (Kulmus 1632), Lorenz Heister (1683–1758) (Heister 1719) and Albrecht von Haller (1708–1777), who inspired Diderot's *Éléments de physiologie* (Diderot 1773–1782, posthumous) (von Haller 1757–1769) (Diderot 2004[12]). French anatomists Exupere-Joseph Bertin (1712–1781) (Bertin 1748), Joseph-Guichard Duverney (1648–1730) (Duverney 1683), Jean-Baptiste Senac (1693–1770) (Senac 1749) and Tarin (1750, 1753a, 1753b) are also in the work. However, the *Dictionnaire raisonné* did not report anything original or recent in the field of human representation in the mid-18th century. Given that the volumes of the *Anatomie* plates appeared in 1765, the *Encyclopédie*, with the exception of Albinus, offered its readers a "classical" image of man, still the *Vitruvian Man*, inherited from the Renaissance and the classical age, where finalism and the "persistence of life" prevailed over any kind of mechanical disillusionment[13].

In conclusion, the medical companies that followed, beginning in the second half of the 18th century, based on the new vitalism of the Montpellier School (Rey 2000), which was well represented in the work of Diderot (2004), would change almost nothing in the *form of figuration* or *ambiguous* representation of the male (and female) body, the latter being both finalistic and scientific during the Enlightenment. Life, according to the "federal" theory of the organic body proposed by Théophile de Bordeu (de Bordeu 1751, 1818[14]), remained in the mechanical analysis of separate parts rather than in the entirety of the human figure.

---

12 The text explores the various discoveries of the century around the "physical and moral man".

13 "This persistence of life in the figure of the flayed man could be explained by hesitations in biology at the end of the 17th century, and throughout the 18th century. As evidenced by the scandal that swept across Europe following the publication of Julien Offray de La Mettrie's *L'Homme-machine* (La Mettrie 1747), the Cartesian model did not establish exclusively, but generated numerous, often contradictory theories in reaction to it. These contradictions contributed to the persistence of arts and symbolism that were foreign to science in the study of anatomy. The body does not allow itself to be disenchanted without resistance" (Cuir 2019, pp. 53–54).

14 De Bordeu's *Recherches anatomiques sur les glandes* marked the solid affirmation of the organicist "vitalism" of the Montpellier School in the 18th century; however, the form of representation of the body did not stray far from that of the *Encyclopédie*, which was still ambiguous. Canguilhem (1971, pp. 101–116) emphasized the close historical and epistemological relationship between the concept of the machine and that of the living, or "vital", organism, with its characteristics of autonomy, self-regulation, self-preservation, etc. ("Machine and Organism").

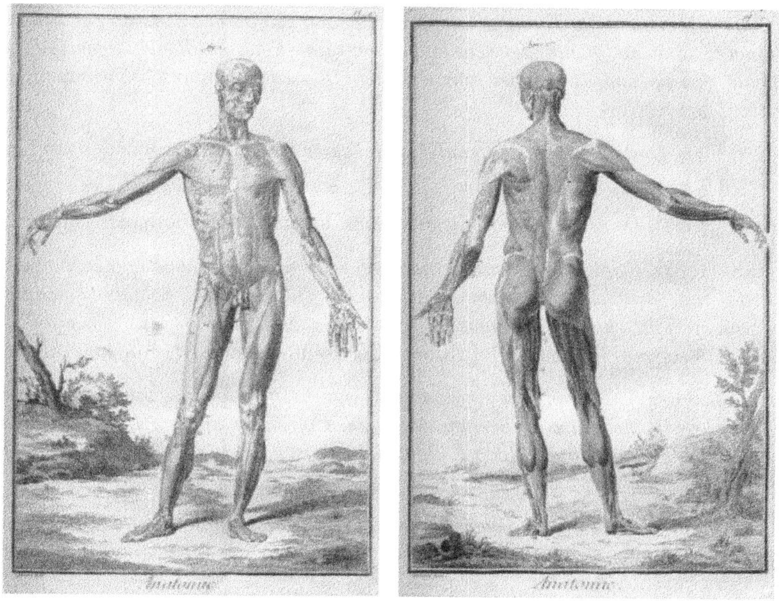

a) b)

Figure 4.31. *Encyclopédie, "Anatomie", plates IV and V (taken from Albinus)*

4.6. References

Primary sources

d'Acquapendente, G.F. (1603). *De venarum ostiolis*. Laurentij Pasquati, Padua.

d'Acquapendente, G.F. (1623). *Opera chirurgica*. Impensis Francisci Bolzettæ, Padua

d'Acquapendente, G.F. (1909). *Tabulae Anatomicae Pictae*. Bibliotheca Marciana, Venice.

Albinus, S. (1747). *Bernardi Siegfried Albini Tabulae sceleti et musculorum corporis humani*. Lugduni Batavorum/Joannem & Hermannum Verbeek, London.

Berrettini da Cortona, P. (1741). *Tabulae Anatomicae*. Typographia Antonii de Rubeis, Rome [Original published 1618].

Bertin, E.-J. (1748). *Lettres sur le nouveau système de la voix, et sur les artères lymphatiques*. Paris.

Bidloo, G. (1685). *Anatomia Humani Corporis, Centum & quinque Tabulis per artificiosiss. G. De Lairesse ad vivum delineatis, Demontrata, Veterum Recentiorumque Inventis explicata plurimisque, hactenus non detectis, Illustrata*. Sumptibus Viduæ Joannis a Someren, Amsterdam.

Bidloo, G. (1690). *Ontleding des menschelyken lichaams*. Weduwe van Joannes van Someren, Amsterdam.

Blankaart, S. (1678). *De nieuw hervormde anatomia*. Johannes ten Hoorn, Amsterdam.

Blankaart, S. (1687). *Stephani Blancardi Anatomia reformata, sive concinna corporis humani dissectio, ad neotericorum mentem adornata, plurimisque tabulis calcographicis illustrata. Accedit ejusdem authoris de balsamatione, nova methodus a nemine antehac hoc modo descripta*. Jordaanum Luchtmans/Cornelium Boutesteyn, Lugduni Batavorum.

de Bordeu, T. (1751). *Recherches anatomiques sur la position des glandes et sur leurs actions, par M. Théophile de Bordeu, Médecin, Conseiller du Roi, Inspecteur des Eaux Minérales de la Généralité d'Auch & Pau, & Professeur d'Anatomie à Pau*. G.F. Quillau, Paris.

de Bordeu, T. (1818). *Œuvres complètes, précédées d'une notice sur sa vie et sur ses ouvrages*, 2 volumes. Caille et Ravier, Paris.

Bourdon, A. (1678). *Nouvelles tables anatomiques où sont représentée au naturel toutes les parties du corps humain*. Laurent d'Houry, Cambray/Paris.

Canano, G. (1541). *Musculorum humani corporis picturata dissectio*. Francesco de' Rossi il giovane, Ferrara.

Casseri, G.C. (1600). *De vocis auditusque organis historia anatomica*. Excudebat Victorius Baldinus, Ferrara.

Casseri, G.C. (1627). *Iulii Casseri Placentini Tabulae anatomicae*. Evangelista Deuchinus, Venice.

Chelseden, W. (1713). *The Anatomy of the Human Body. Illustrated with Twenty-Three Copper-Plates of the Most Considerable Parts; All Done After the Life*. N. Cliff/ D. Jackson, London.

Chelseden, W. (1733). *Osteographia, or the Anatomy of the Bones*. William Bowyer, London.

Colombo, M.R. (1559). *De re anatomica*. Nicolai Bevilacquæ, Venice.

Descartes, R. (1996). *L'Homme*. In *Œuvres*, volume VI, Adam, C., Tannery, P. (eds). Vrin, Paris [Original published 1662/1664].

Diderot, D. (1749). *Lettre sur les aveugles, à l'usage de ceux qui voient*. Paris.

Diderot, D. (1751–1772). *Encyclopédie, ou Dictionnaire Raisonné des Sciences, des Arts et des Métiers*, 17 volumes. Le Breton/Briasson/Durand, Paris.

Diderot, D. (2004). *Éléments de physiologie, en Appendice le Manuscrit de Pétersbourg.* Texte établi, présenté et commenté par P. Quintili. Honoré Champion, Paris.

Dionis, P. (1690). *L'Anatomie de l'Homme, suivant la Circulation du Sang, et les nouvelles Découvertes. Démontrée au Jardin Royal par Mr Dionis, premier Chirurgien de feuës Mesdames les Dauphines, & Maître Chirurgien Juré à Paris. Cinquième édition Exactement revuë & beaucoup augmentée par l'Auteur ; avec une ample Dissertation sur la génération, & plusieurs explications de faits particuliers, accompagnez de Figures nouvelles.* Laurent d'Houry, Paris.

Dionis, P. (1707). *Cours d'opérations de chirurgie, démontrées au Jardin royal.* Laurent d'Houry, Paris.

Drake, J. (1707). *Anthropologia Nova, or, a New System of Anatomy.* S. Smith & B.J. Walford, London.

Duverney, J.-G. (1683). *Traité de l'organe de l'ouïe, contenant la structure, les Usages & les Maladies de toutes les parties de l'Oreille. Par M. Du Verney, de l'Académie Royale des Sciences, Conseiller, Médecin Ordinaire du Roy, & Professeur en Anatomie & en Chirurgie au Jardin Royal des Plantes.* Chez Estienne Michallet, Paris.

Eustachi, B. (1714). *Tabulae anatomicae clarissimi viri Bartholomæi Eustachii, quas e tenebris tandem vindicatas et Clementis XI Pontifici Maximi Munificentia dono acceptas, Præfatione, Notisque illustraviv, ac ipso suæ Bibliothecæ dedicationis die publici juris fecit Joh. Maria Lancisius, Intimus Cubicularius, et Achiater Pontificius.* Ex Officina Typographica Francisci Gonzagæ, Rome.

de Graaf, R. (1668). *De virorum organis gerationi inservientibus, de clisteribus et de usu siphonis in anatomia.* Lugduni Batavorum, Ex Officina Kackiana.

Guidi, G. (1611). *De anatome corporis humani.* Iuntas, Venice.

von Haller, A. (1757–1769). *Elementa physiologiæ corporis humani*, 8 volumes. Bousquet, Lausanne.

Hartsoeker, N. (1694). *Essay de Dioptrique.* Jean Anisson, Paris.

Harvey, W. (1628). *Exercitation anatomica de motu cordis et sanguinis in animalibus.* Sumptibus Guilielmi Fitzeri, Frankfurt.

Heister, L. (1719). *Compendium anatomicum. Totam rem Anatomicam brevissime complectens.* Editio altera prima longe auctior, Altorfii, in bibliopolio Kohlesiano.

de Ketham, J. (1491). *Fasciculus medicinae.* Johann et Gregorio, Venice.

Kulmus, J.-A. (1732). *Tabulae anatomicae in quibus corporis humani omniumque ejus partium structura & usus brevissime explicantur: accesserunt majoris perspicuitatis causa annotationes et tabulae aeneae auctore Jo. Adamo Kulmo.* Janssonio-Warsbergios, Amsterdam.

de La Mettrie, J.O. (1747). *L'homme machine*. Elie Luzac, Leyde.

Lancisi, G.M. (1728). *De motu cordis, et aneurismatibus, opus postumum in duas partes divisum*. Joannem Mariam Salvioni, Rome.

Le Bovier de Fontenelle, B. (1731). Éloge de M. Ruysch. In *Histoire de l'Académie Royale*. Bibliothèque Royale, Paris.

van Leeuwenhoek, A. (1722). *Antonii à Leeuwenhoek, Regiae societatis Anglicanae socii Opera omnia, seu Arcana naturae, ope exactissimorum microscopiorum detecta, experimentis variis comprobata epistolis, ad varios illustres viros... comprehensa, & quatuor tomis distincta. Editio novissima, prioribus emendatior*. Joh. Arnold. Langerak, Lugduni Batavorum.

van Leeuwenhoek, A. (2014). *The Collected Letters of Antoni van Leeuwenhoek*, volume XVI. Taylor & Francis, London.

Leopardi, G. (2007). *Petites œuvres morales*, J. Gayraud (trans.). Éditions Allia, Paris.

de' Liuzzi, M. (1478). *De omnibus humani corporis interioribus membris Anathomia*. Padua.

Locke, J. (1690). *An Essay Concerning Human Understanding*. Thomas Basset, London.

Malpighi, A. (1661). *De pulmonibus observationes anatomicae*. Typis Jo. Baptistæ Ferronii, Bologna.

Malpighi, A. (1665). *Tetras anatomicarum epistolarum De Lingua, et Cerebro*. Typis HH, Victorij Benatij.

Mascagni, P. (1787). *Vasorum lymphaticorum corporis humani historia et iconographia*. Pazzini Carli, Siena.

Morgagni, G.B. (1706). *Adversaria Anatomica Prima. Ab eo nuper in eadem Academia publice lecta, Multis deinde accessionibus, novisque iconismis adaucta, et viris praestantissimis ejusdem academiae Principi, ac Sodalibus D.D.D.* Typis Ferdinandi Pisarri, Bologna.

Morgagni, G.B. (1761). *De sedibus, et causis morborum per anatomen indagatis libri quinque. Dissectiones, et animadversiones, nunc primum editas complectuntur propemodum innumeras, medicis, chirurgis, anatomicis profuturas. Multiplex praefixus est Index rerum, & nominum accuratissimus*. Remondiniana, Venice.

Morgagni, G.B. (1762). *Adversaria anatomica omnia*. Remondiniana, Venice.

Nuck van Leiden, A. (1690). *Sialographia et ductum aquosorum anatome nova*. P. Vander Aa, Lugduni Batavorum.

Paracelsus (1922). *Sämtliche Werke*, volume 14. K. Sudhoff, Munich/Berlin.

Paracelsus (1955). *Sämtliche Werke*. K. Goldammer, Wiesbaden.

Paracelsus (1983). *Les sept livres de l'Archidoxe Magique*, M. Haven (trans.). Éditions Bussière, Paris.

de Plantade, F. (1699). Extrait d'une lettre contenant une observation microscopique de la semence par M. Dalenpatius. In *Nouvelles de la République des Lettres*. Henry Desbordes, Amsterdam.

Ruysch, F. (1702–1715). *Thesaurus anatomicus secundus [-nonus]*. Parts II-IX. Joannes Wolters, Amsterdam.

Ruysch, F. (1716). *Thesaurus anatomicus primus [-decimus]... Het eerste [-tiende] anatomisch cabinet...* Part I. Jansson-Waesberg, Amsterdam.

Ruysch, F. (1721). *Thesaurus anatomicus primus [-decimus]... Het eerste [-tiende] anatomisch cabinet...* Part X. Jansson-Waesberg, Amsterdam.

de Senac, J.-B. (1749). *Traité de la structure du cœur, de son action et de ses maladies*, 2 volumes. Jacques Vincent, Paris.

van den Spiegel, A. (1626). *De formato foetu, liber singularis, æneis figuris exornatus, epistolæ duæ anatomicæ, tractatus de arthritide/opera posthuma studio Liberalis Cremæ [...] edita*. Jo. Bap. de Martinis & Liuiu[m] Pasquatu[m], Peddie.

van den Spiegel, A. (1627). *De Humani Corporis Fabrica Libri X Tabulis Aere Incisis [...] opus posthumum. Daniel Bucretius [...] jussu authoris in lucem profert*. Impensis & Caelo Matthaei Meriani, Venice.

Tarin, R. (1750). *Adversaria anatomica, de omnibus corporis humani partium, tum descriptionibus, cum picturis. Adversaria anatomica prima, de omnibus cerebri, nervorum & organorum functionibus animalibus inserventium, descriptionibus & iconismis. Autore Petro Tarin*, Joannis Francisci Moreau, Paris.

Tarin, R. (1753a). *Ostéo-graphie, ou description des os de l'adulte, du fœtus, &c. Précédée d'une introduction à l'étude des parties solides du corps humain*. Briasson, Paris.

Tarin, R. (1753b). *Dictionnaire anatomique, suivi d'une bibliothèque anatomique et physiologique*. Briasson, Paris.

Varolio, C. (1573). *De nervis opticis nonnulisque aliis praeter commune opinione in humano capite observatis*. Paulum & Antonium Meiettos fratres, Padua.

Varolio, C. (1591). *Anatomiae sive de resolutione corporis humani ad Caesarem Mediovillanum libri IV*. Apud Ioannem Wechelum & Petrum Fischerum consortes, Frankfurt.

Vesalius, A. (1543). *De humani corporis fabrica*. Per Ioannem Oporinum, Basel.

Vieussens, R. (1716). *Raymundi Vieussens Doctoris Medici Monspeliensis Neurographia Universalis. Hoc Est, Omnium Corporis Humani Nervorum simul & cerebri, medullaeque spinalis Descriptio Anatomica ; eaque integra et accurata, variis iconibus fideliter & ad vivum delineatis, aereque incisis illustrata : cum ipsorum actione et usu, Physico discursu explicatis, Editio Novissima*. Joannem Certe, Lugduni Batavorum.

da Vinci, L. (1510). Codex Windsor. In *Anatomical Drawings from the Royal Library Windsor Castel.* The Metropolitan Museum of Art, New York.

da Vinci, L. (2014). *Tous les dessins*. Taschen, Paris.

Willis, T. (1664). *Cerebri anatome: cui accessit, nervorum descriptio et usus*. Gerbrandum Schagen, Amsterdam.

Secondary sources

Adelmann, H.B. (1966). *Marcello Malpighi and the Evolution of Embryology*, 5 volumes. Ithaca, Cornell.

Canguilhem, G. (1971). *La connaissance de la vie*. Vrin, Paris.

Carlino, A. (1994). *La fabbrica del corpo. Libri e dissezione nel Rinascimento*. Einaudi, Turin.

Carlino, A. (1998). Vésale retombé sur terre. De la culture visuelle de l'anatomie à la Renaissance. In *Corps à vif. Artistes et anatomie*, Carlino, A., Petherbridge, D., Ritschard, C. (eds). Musée d'Art et d'Histoire, Geneva.

Carlino, A. (1999). L'impératif métaphorique : quelques réflexions autour de l'illustration anatomique : XVe-XVIIIe siècle. *Traverse. Zeitschrift für Geschichte*, VI(3), 23–24.

Carlino, A. (2010). *L'anatomia tra arte e medicina. Lo studio del corpo nel tardo Rinascimento*. Silvana Editoriale, Milan.

Carlino, A. and Ciardi, R. (2012). *Visioni anatomiche. Le forme del corpo negli anni del Barocco*. Silvana Editoriale, Milan.

Carlino, A., Petherbridge, D., Ritschard C. (1998). *Corps à vif. Artistes et anatomie*. Musée d'Art et d'Histoire, Geneva.

Cherici, C. and Dupont, J.-C. (2008). *Les querelles du cerveau. Comment furent inventées les neurosciences*. Vuibert, Paris.

Cherici, C. and Dupont, J.-C. (2019). *L'anatomie sans les arts ? Le corps en images à l'époque moderne (XVIe-XVIIIe siècles)*. Hermann, Paris.

Cosmacini, G. (1997). *L'arte lunga. Storia della medicina dall'Antichità a oggi*. Laterza, Rome/Bari.

Croft, W.J. (2006). *Under the Microscope. A Brief History of Microscopy*. World Scientific Publishing Company, London.

Cuir, R. (2019). De l'écorché animé au cadavre, vers la spécialisation du savoir anatomique, In *L'anatomie sans les arts ? Le corps en images à l'époque moderne (XVIe-XVIIIe siècles)*, Cherici, C. and Dupont, J.-C. (eds). Hermann, Paris.

Deleuze, G. (1988). *Le Pli. Leibniz et le Baroque*. Éditions de Minuit, Paris.

Elkins, J. (1986). Two conceptions of the human form: Bernhard Siegfried Albinus and Andreas Vesalius. *Artibus et historiae*, 14(7), 91–106.

Engelhardt, E. (2021). Marcello Malpighi: The nervous system under microscope. *Arquivos de Neuro-Psiquiatria (Historical Notes)*, 79(4). doi: 10.1590/0004-282X-ANP-2020-0309.

Fortuna, F. (2014). Le *Tabulæ anatomicæ* di Bartolomeo Eustachio. *Journal of Italian Medical Education*, 64, 2913–2916.

Foucault, M. (2001). *L'herméneutique du sujet, cours au Collège de France, 1981-1982*. Gallimard/Le Seuil, Paris.

Ghosh, S., Sharma, K., Biswas, S., Chakraborty, S. (2014). Adriaan van den Spiegel (1578–1625): Anatomist, physician, and botanist. *Clinical Anatomy*, 27(7), 952–957. doi: 10.1002/ca.22414.

Grmek, M.D. (2014). *Histoire de la pensée médicale en Occident. Tome 2 : de la Renaissance aux Lumières*. Le Seuil, Paris.

Gros, F. (2019). L'anatomie, l'art et le corps en images à l'époque moderne. In *L'anatomie sans les arts ? Le corps en images à l'époque moderne (XVIe-XVIIIe siècles)*, Cherici, C. and Dupont, J.-C. (eds). Hermann, Paris.

Guest, C.E.L. (2012). *Anatomy and the Baroque: Frederik Ruysch, Thesaurus Anatomicus 1709-26* [Online]. Available at: https://www.academia.edu/30531363/Anatomy_and_the_Baroque_Frederik_Ruysch_Thesaurus_Anatomicus_1709_26_Edward_Worth_Library_Dublin_Oct_2012.

Hansen, J.V. (1996). Resurrecting death: Anatomical art in the cabinet of Dr. Frederik Ruysch. *The Art Bulletin*, 78(4), 663–679.

Kooijmans, L. (2011). *Death Defied. The Anatomy Lessons of Frederik Ruysch*, D. Webb (trans.). Brill, Leiden/Boston.

Le Nen, D. and Laulan J. (2010). *La main de Léonard de Vinci*. Springer, Paris.

Maienschein, J. (2014). *Embryos Under the Microscope. The Diverging Meanings of Life*. Harvard University Press, Cambridge.

Persaud, T.V.N., Loukas, M., Tubbs, R.S. (2014). *A History of Human Anatomy*, 2nd edition. Charles C. Thomas, Springfield/Illinois.

Quintili, P. (2008). Les "Éléments de physiologie" et l'histoire du cerveau. Diderot face à la physiologie de son temps. In *Les querelles du cerveau. Comment furent inventées les neurosciences*, Cherici, C. and Dupont, J.-C. (eds). Vuibert, Paris.

Quintili, P. (2019). La question de l'image du corps en Occident moderne (XVIIe et XVIIIe siècles). In *L'anatomie sans les arts ? Le corps en images à l'époque moderne (XVIe-XVIIIe siècles)*, Cherici, C. and Dupont, J.-C. (eds). Hermann, Paris.

Rey, R. (2000). *Naissance et développement du vitalisme en France de la deuxième moitié du 18e siècle à la fin du Premier Empire*. Voltaire Foundation, Oxford.

Rossi, P. (2000). Il fascino della magia e l'immagine della scienza. In *Storia della scienza moderna e contemporanea*, volume 1. TEA, Milan.

Snyder, L.J. (2015). *Eye of Beholder. Johannes Vermeer, Antony van Leeuwehoek and the Reinvention of Seeing*. Head of Zeus, London.

Solinas, G. (1967). *Il microscopio e le metafisiche. Epigenesi e preformismo da Cartesio a Kant*. Feltrinelli, Milan.

Tubbs, R.S., Shoja, M.M., Loukas, M., Agutter, P. (2019). *History of Anatomy. An International Perspective*. Wiley Blackwell, New York.

5

The Observation and Classification of Living Beings: The Animal, 16th–18th Centuries

Jean-Luc GUICHET
CERCLL, Université de Picardie Jules Verne, Amiens, France

5.1. Introduction

The observation and classification – or at least categorization – of animals date back to ancient times, even prehistoric times, as soon as language and representation appeared. Cave art focuses almost exclusively on animal forms and paints vast frescoes, grouping them into easily identifiable ensembles, such as groups of bison and lionesses. The realism of these representations, attention to detail, mannerisms and typical movements shows an extremely careful observation that gives us the illusory feeling of perceiving reality as it was. On the basis of these examples, one might think that the distribution of animal species has always been self-evident to a human mind that has only had to record a natural and immediately visible geography of living things. However, regardless of the observational qualities that are often revealed by the oldest representations, at least those that attest to direct knowledge of the animal, it should not be assumed that the overall representation of the living world necessarily followed the same reality that we naturally tend to project onto these depictions. In real terms, our classification of animal populations, grouping them according to their anatomical and phylogenetic affinities into hierarchical

History of Life Sciences and Medicine 2,
coordinated by Céline CHERICI.

groups (in ascending order: species, genus, family, order, class, phylum, kingdom), which is seemingly obvious to us in retrospect, is the result of historical work that has structurally and culturally informed our perspective. This scientific work of classification and of establishing the anatomical, physiological and behavioral profiles of the main species, which we have inherited, actually began during the Renaissance and reached its first culmination in the 18th century, laying the foundation for the decisive progress made in the 19th and 20th centuries.

5.2. Ancient and medieval legacy

5.2.1. *Aristotle*

Aristotle's hugely impactful work cannot be overlooked. The latter was, in fact, the first scientific work on animals. The first major idea – a positive one – was simply that within the abundance of animals, a global order of living beings could be identified. However, Aristotle does not accompany this guiding principle in his writings with a system for identifying animals according to a structured and indexed classification. It is quite possible, as reviewers have pointed out, that this relative indeterminacy stems from the great complexity of this undertaking[1], a problem that was passed on to subsequent naturalists. The second major idea – this time of a metaphysical nature – was also decisive for subsequent eras, despite its allusive nature in Aristotle's text itself: that of a scale of beings (*scala naurae*). Despite how it seemed, the two ideas did not always unify in the centuries that followed, as the work of classification often artificially strove to adhere to this assumption of perfect continuity. However, this idea of a ladder or chain of beings fluctuated between two extremes: a continuity without disruption, where one passes imperceptibly from one form to another, and a hierarchy of species according to a graded series, from the least perfect to the most perfect, which became established in Christianity and would raise many problems. In any case, Aristotle's influence, like a guiding thread, constantly presupposed an order in the empirical world of living beings, regardless of how colorful and disconcerting it was, which would become increasingly so with the proliferation of observations and travels.

This idea of order, as well as that of unbroken continuity, or even a hierarchy, linking all creatures, flourished during the Renaissance and until the 18th century.

1 "[...] it follows that Aristotle's zoological work [...] contains absolutely nothing that resembles a complete and definitive zoological classification" (Daudin 1983, p. 18). However, we might add the following: "What we can say, however, is that Aristotle sensed the full complexity of the classification of living beings" (Petit and Théodoridès 1962, p. 84).

5.2.2. *The legacy of late antiquity and the Middle Ages*

Antiquity cannot be summed up by Aristotle alone, and toward the end of this period, a rich discourse on animals developed, far removed from the rigorous requirements of the Stagirite and, to say the least, very receptive to fantastic stories. Among these authors, Pliny the Elder (23–79) and, to a lesser extent, Dioscorides (between 20 and 40–c. 90) and Aelian (c. 175–c. 235), exerted a significant and lasting influence. A bestiary from the second century, *Physiologus* and the *Etymologies* of Isidore of Seville in the seventh century also had a huge impact.

In this type of work, alongside reliable observations (e.g. in Galen, 129–c. 201) or critical distancing (as occasionally found in Isidore of Seville), animals often became the mediums for a world of imagination that attributed fantastical and often supernatural qualities to them. This literature, and the medieval texts inspired by it, reinterpreted and enriched ancient heritage through the prism of symbolism that was specific to Christianity. Animals fed the exemplary catalog of vices and virtues that believers had to follow with complete clarity; thus, the pelican sometimes symbolized parental devotion and, at other times, the sacrifice of Christ, with its nurturing behavior of regurgitating food directly into the throats of its young being interpreted as the gift of its entrails to its offspring or of its blood to all of humanity. It is understandable that, since interest in this literature is not focused on the objective reality of animals but on the symbolism they embody, the categorization of living beings follows rules that have nothing to do with scientific classification[2].

The Middle Ages, in its own realm, saturated with religiousness, viewed animals from the perspective of its main concerns – the Christian conception of the world and the salvation of the individual – and demanded to be understood as much as possible from this perspective and not another. However, from then on, it developed a very powerful imaginary world, mixing ancient references and Christian marvels, which lasted until the 16th century and even beyond, as exemplified in a page by Buffon where, celebrating Pliny and his grand style, he embraces both the sense of modesty and the embryonic religion that Pliny attributed to the elephant. Furthermore, in the Arab world, interest in animals, in keeping with Antiquity, developed and gave rise to studies that were often already very advanced in terms of species, habits and anatomy.

2 This is what Foucault points out, for example, in *Les mots et les choses*. However, here again, there are exceptions: for example, Albert the Great (1200–1280) and Emperor Frederick II of Hohenstaufen (1194–1250) are characterized by their concern for accurate observation and causal explanation.

5.3. The scholarly 16th century: a century of inventory and transition

The Renaissance marked a significant shift. Admittedly, this was contradictory: on the one hand, direct observation became increasingly important, but, on the other hand, references to Antiquity, especially to Aristotle, remained almost systematic. The 16th century thus appears overall as a scholarly century, devoted to the effort to rediscover literary heritage through the living world observed or, more often, reported in travelers' accounts. What was *seen* certainly grew rapidly, owing to major expeditions, as well as to a developing spirit of accuracy, but it was often necessary to justify oneself against the impact of something that was *read*, which was still seen as more credible, even if, as the century progressed, it was increasingly challenged. Notably, *a contrario*, references to Aristotle (e.g. by Edward Wotton, author of *De differentis animalium* in 1552), often served as a safeguard of rationality against the many falsehoods that were circulating. Furthermore, the effort to create a classification became the main focus and unifying force of scientific research, even though no significant progress was made in the immediate present[3].

5.3.1. *Classification inadequacy*

The collection of new species reported from different corners of the world is increasingly evident, exceeding ancient heritage and making it imperative to find clear and consensual classification principles. However, a major obstacle to this project is the power of strong ideas: persistent respect for ancient references, the principle of continuity and, generally, hierarchy, in addition to a declared anthropocentrism justified on religious grounds. Another obstacle stems from practical concerns, such as the food, pharmaceutical and medical uses of animal substances. An indication of classification embarrassment resulting from this combination of data overload, inadequate categories and overlapping concerns is the frequent use of alphabetical order as a default. Another common expedient is to classify animals according to their environment: all those that live in water are grouped under the title *fish*, whether they are true fish or cetaceans, pinnipeds, etc., not to mention a few unicorns and mermaids; those that fly are grouped together under the name *birds*, including true birds and bats; those that crawl are grouped

3 "[...] from the end of the 16th century to the end of the 18th century, the project of distributing all living beings, animals or plants, into a hierarchy of collective units, that were rigorously included within one another, gained such a sway over the mind of naturalists that it eventually appeared to them as the very formula of their scientific task [...]" (Daudin 1983, p. 20).

under the name *insects* and *worms*, in addition to all invertebrates, which concern many species, including anurans (frogs, toads) and urodela (newts, salamanders), that are found in several categories at once.

5.3.2. *Influence of the principle of continuity: the problem of intermediaries and the hierarchy of perfection*

Such overlaps like the latter were not necessarily perceived as a flaw. The widespread desire to verify the continuity of the chain of beings justified the admission of intermediate or "ambiguous" beings, which provided a link either between kingdoms – sponges, for example, fulfilling the role of zoophytes and linking animals and plants – or within groups themselves, with bats, for example, being situated between birds and land animals. This assumed perfection often hindered observation: bats must remain half bird and half something else to fulfill their role as a link. Since the ambiguity of certain beings was considered necessary for the continuity of creation, sharp distinctions represented a potential threat in this regard. The problem was further complicated when this continuity was interpreted, as is most often the case, as a scale of perfection: How can species not only be ordered but also according to a linear and graded scale from least perfect to most perfect? What methodological and axiological criteria should be used to achieve this? Here, human beings have become a reference point: assumed to be the most perfect of animals, they serve as a standard for other species. However, we can clearly see the difficulties in establishing these criteria and applying them satisfactorily to the growing diversity of the animal world that has been cataloged. Furthermore, certain points in this chain of beings raise vital problems that this idea will never be able to resolve: How, specifically, can we conceive of a continuity between finite creatures and the infinite God?

5.3.3. *Specific challenges in zoology*

Another set of difficulties arose from practical conditions that almost always require studying living organisms through dead organisms. Unlike in botany, animals are mobile and sometimes dangerous, making them difficult to approach. In addition, a major problem was preservation, as techniques were not yet fully developed: apart from a few rare living animals, only remains or hard parts (shells, carapaces, teeth, bones, etc.) or dried parts, either naturally or by desiccation, were generally reported. The specimens that scientists acquired were therefore fragmented or altered, leading to misinterpretations, especially as there were many fakes (more or less successful, for example, the so-called mermaids combining the bodies of fish and fetuses or the thoraxes of small monkeys), and a whole trade was developed. The problem was exacerbated by the fact that, unlike plants, animals do not carry their organs on their bodies. The visibility of their anatomical structure was therefore lost to the observer, even though this was valuable information for determining criteria for identifying and classifying species according to their true affinities and differences.

Faced with these difficulties, the practice of traveling increasingly competed with books or at least complemented and corrected them. The rise of natural history cabinets and, above all, the proliferation of the first menageries and the development of anatomy, particularly comparative anatomy, even though it remained in its infancy, allowed for more direct and regular observation of real animals.

5.3.4. *Cabinet scholars and naturalist travelers*

The Renaissance thus saw true developments in the writings on animals, with a large number of authors. These can be broadly grouped into two categories, although they often belong to both categories: on the one hand, scholars who remained more attached to their libraries (although they sometimes practiced observation) and often developed systems – with a classificatory or explanatory ambition – while discussing ancient authors and, on the other hand, naturalist travelers (often also highly literate). The latter were generally physicians, as there were close links between medicine and zoology, owing to the therapeutic use of animal parts or products (e.g. the famous castoreum) but also with anatomy, as dissection of the human body remained forbidden until the 13th century (when it was first practiced in Italy, then in France in the 14th century), making it necessary to study the animal body instead (often pigs, considered close to humans). There was also a meeting of minds between scholars and amateurs from a wide variety of backgrounds, as scientific specialization was far from the norm at the time and considerations about animals were very common in all kinds of writings. This boom in zoological studies particularly fostered interest in ichthyology and, to a lesser extent, ornithology. Insects and invertebrates, however, remained largely neglected until the following century.

Among all these authors, the following stand out: Conrad Gessner (1516–1565), Pierre Belon (1517–1564) and Guillaume Rondelet (1507–1566), the former more as a cabinet scholar (although he also traveled extensively throughout Europe) and the latter two more as naturalist travelers (although they were also scholars). These three authors, who knew each other, combined extensive zoological knowledge with a commitment to observation and critical rigor that was significantly greater than before, coupled with regular dissection practices. This was certainly not without lingering beliefs at times (such as, among other things, the spontaneous generation of certain animals or the generation of barnacle geese from certain trees) or even outright gullibility (such as the belief that the remora fish could single-handedly stop the largest ships) and frequent confusion between species, but overall, they stood out more than other naturalists of the time who shared these traits. In addition, the illustrations in their writings are often characterized by rather unusual objectivity, making them true naturalistic drawings.

The nickname "German Pliny", once given to Gessner (who was actually Swiss), is indicative of the reputation he had. His immense body of work covers the entire animal world known at the time, with his main work being *Historiae animalium* (1551–1587). This genuine compendium demonstrates a remarkable desire to synthesize the knowledge of the time and to provide exhaustive information on each species, each of which is the subject of a monograph. This ambition, however, certainly contributed to the choice of a simple alphabetical order, effectively renouncing any real classification, which was apparently considered too complex and improbable. Although the work still includes a number of fabulous animals, it seems to be due to the desire to provide as complete an inventory as possible, which was the motive behind the work. Gessner expressed many doubts about them and, in general, always took care to verify the information he reported, either through observation or by cross-referencing with other sources.

Belon and Rondelet both left behind a body of work dominated by ichthyology and, in the case of the former, also ornithology. Their work reveals a keen spirit of observation and a seemingly critical attitude toward ancient and popular sources. Belon traveled throughout the Mediterranean, particularly in the Middle East, and is best known for *L'Histoire naturelle des estranges poissons marins* (Belon 1551) and *L'Histoire de la nature des oyseaux* (Belon 1555a), although he dealt with the entire animal world, introducing many new species. Despite his often chaotic classification, his considerable work demonstrates a methodical sense of observation, with detailed anatomical descriptions, comparisons between organisms and a very systematic examination of stomach contents, enabling him to identify diets. For Rondelet, he was a more modest traveler (having mainly traveled in Italy and along the Atlantic coast) and an author of a major work, *Libri de piscibus marinis in quibus verae piscium effigies expressae sunt*, which was split into two volumes, in 1554 and 1555, and brought together in a French edition in 1558, *L'Histoire entière des poissons* (The Complete History of Fish). Rondelet stood out for his meticulous observation, the range of species he recorded and the remarkable precision of the illustrations (by an unknown artist) in his works.

Other works worth mentioning are those of Gyllius (Pierre Gilles, 1490–1555), Olaus Magnus (1490–1557), Edward Wotton (1492–1555), Hippolyte Salviani (1514–1572), Ulisse Aldrovandi (1522–1605), Carolus Clusius (Charles de l'Écluse, 1526–1609), Fabrice d'Acquapendente (1533–1619), the Bauhin brothers (Jehan, 1541–1612, and Caspar, 1560–1624), Thomas Muffet (1552–1604), as well as Leonardo da Vinci (1452–1519), Ambroise Paré (1510–1590) and Andreas Vesalius (1514–1564), to name the best known.

5.4. The 17th century: disruption at work

Just as zoology owed much of its impetus in the 16th century to the expansion of the world owing to great expeditions (which continued over the following centuries), so did the very rich 17th century, which gave it its name (by Sperling, 1603–1658), owing much of its growth to decisive technical innovations, specifically the development of the microscope. The number of new species discovered at the time thus increased considerably. As a result, the need for an effective classification system became ever more acute. In addition, the decisive advances in anatomy are linked to the practice of dissection, which has become almost systematic. The scientific world began to organize itself, and owing to the proliferation of academies and learned societies, it no longer depended solely on the goodwill of patrons. All these advances, which pointed to an ever-increasing role for direct observation, encouraged a progressive step further from the influence of the Elders.

The history of zoology in this century has been driven by strong, highly active and innovative personalities, working either individually and independently (while maintaining close correspondence with many other scholars) or surrounded by a "team" such as in Rome with the Academia dei Lincèi (the Academy of the Lynx, so called, among other things, in reference to the magnified vision made possible by the microscope), the oldest in Europe or the Académie des sciences in Paris, the Royal Society in London, etc. The former were often primarily discoverers, whereas the latter were more generally systematists and classifiers, with, of course, a great deal of overlap between the two groups.

5.4.1. *The age of discovery and the microscope*

The contribution of the microscope is immeasurable: it not only opened up a new field of life, that of previously unknown microorganisms but also provided a new approach to already known species, in which observers were now rediscovering. By inspecting organs and tissues at a scale that was previously inaccessible to the human eye, organisms could reveal their close structures, with the “micro” explaining the “macro”. Scientists were mainly Italian – Francesco Stelluti (1577–1646), known in particular for his work on bees, and Marcello Malpighi (1628–1694), considered the founder of microscopic anatomy and whose treatise on silkworms, *De Bombyce* (Malpighi 1669), the first detailed monograph on an invertebrate, is a benchmark work – English (Robert Hooke, 1635–1703) and Dutch (Jan Swammerdam, 1637–1680 and Antoni van Leeuwenhoek, 1632–1723). The latter two scholars were individuals who conducted their own research far from any defined collective framework. However, their work was carried out in very different spirits. Van Leeuwenhoek, whose isolation was relative, since he corresponded with the Royal Society and sent them his work throughout his career, proceeded without any discernible method, like an empirical discoverer jumping from subject to subject without bothering to connect the discontinuity of his interests with an explicit line of coherence. Emblematically associated with the microscope (whose history actually spans the entire century), it was above all his ability to exploit it that constituted his major contribution to the science of his time. Van Leeuwenhoek discovered protozoa, studied insects, detailed their anatomy, shed light on their modes of reproduction, and did the same for crustaceans, etc. As for Swammerdam, he constantly complemented his exploratory work with a methodical and scrupulous concern for systematization. In addition, he was highly innovative and perfected techniques, particularly microscopic anatomy, which enabled him to make specific advances in the study of insects.

Thus, observation not only benefited from this magnified vision but also, in a related way, from the practice of dissection (which developed anatomy and embryology in particular) and experimentation (which developed physiology). Conversely, the combination of microscopy and dissection gave significant momentum to entomology and the study of invertebrates in general, a momentum that was more or less contributed to by all the aforementioned microscopists.

5.4.2. *The contributions of dissection and experimentation*

Following the example of the English physician William Harvey (1578–1657) and his use of vivisection (which enabled him to demonstrate blood circulation in 1628 by sacrificing deer and fallow deer in Charles I’s park, which were entirely at his disposal), research on anatomy and physiology in this century focused heavily on animals. Whether studied for their own sake or, more commonly, with a view to understanding human organisms, the observation and understanding of animal organisms have benefited greatly. We will focus on two central figures: the French physician and anatomist Claude Perrault (who was not just the architect of the colonnade of the Louvre) and the Italian physiologist Francesco Redi.

The work of Claude Perrault (1613–1688) cannot be understood without reference to the context of the newly founded Académie des sciences (1666), which he was a member of and where work was regularly shared, thus benefiting from healthy mutual criticism. Another advantage of this institution was its access to the Royal Menagerie of Versailles and the opportunity to dissect animals *postmortem*, most of which were exotic and previously known only by their appearance. Thus, Perrault, along with Duverney and other academics, nicknamed the Parisians, dissected numerous species: elephants (in the presence of Louis XIV, a sign of the role that the natural sciences had acquired at that time, in the eyes of the monarch, as protector of the arts and sciences), ostriches, crocodiles, cassowaries, pangolins, etc., and even a camel, the last episode causing him fatal sepsis. His main work, *La mécanique des animaux* (Perrault 1680), was a significant contribution: drawing on his wealth of observations, he developed a rigorous and scientific comparative anatomy, thus initiating a more synthetic and comprehensive understanding of organisms, which would form the basis for future progress.

Francesco Redi (1626–1697), known for his experimental rejection of spontaneous generation, was also a zoologist and, in particular, produced precise and methodical work on parasites in the most diverse animal species (including invertebrates), work that was taken on and furthered by his students, including Vallisnieri, and opened up a new field of scientific knowledge. We should also mention John Jonston (1603–1675), Thomas Willis (1621–1675), Nicolas Sténon (Niels Steensen, 1638–1686), Martin Lister (1638–1712), Reinier de Graaf (1641–1673) and Edward Tyson, whom we discuss further below.

This progression of the observational, verifying and experimental gaze produced a powerful global rationalizing effect on the knowledge of living beings, increasingly discrediting popular beliefs and the legacy of bestiaries and ancient literature (notably Pliny) in informed circles. However, resistance did not disappear easily: in particular, the idea of spontaneous generation persisted not only in the minds of the public but also in scholarly circles (albeit limited to small organisms).

However, all these factors, combined with the yield reported by major expeditions, made it increasingly necessary and urgent to develop a classification system that reflected these new data.

5.4.3. *The classification crisis*

The 17th century certainly saw a wealth of discoveries thanks to continued travel, increased observation through the microscope, further study and refinement of anatomy, advances in techniques (notably the injection of solidifying or coloring substances to reveal blood vessels and other structures), improvements in conservation methods thanks to alcoholic solutions and increasing precision in scientific drawing, among other things. However, overwhelmed by this wealth of knowledge and under the weight of its own success, the century seemingly faced a kind of classification crisis. The major challenge for scholarly work, centered on this crisis, was to overcome it, or at least to begin doing so. One possibility, however, emerged: this reframing of research through anatomy and the microscope made it possible to base classification on internal characteristics, rather than solely on external, morphological or even environmental factors (classification of species by their places of existence), or even alphabetical order, bringing it closer to the "affinities" of the "natural method", an ideal that many people had hoped for. Classification and observation now intersect and support one another at a deeper and more fundamental level.

5.4.4. *John Ray's classification*

This classification problem began to be overcome because of the significant contribution of John Ray, an English naturalist, botanist and zoologist (like many scholars of that time, before disciplines were divided into separate fields).

The essential contribution of John Ray (1627–1705), assisted by Francis Willughby (1635–1672), who died prematurely and whose specific contribution is difficult to determine, was to establish a classification on the basis of anatomical criteria. On this basis, the concept of species, which had previously been rather vague and variable, was now anatomically, or structurally, delineated and defined as an invariant across successive generations, thereby acquiring scientific consistency and clarity (Carus 1880, pp. 339–340). For example, the category of fish usually included all animals living in water and therefore, in addition to true fish, included mammals (delphinids and cetaceans), crustaceans, reptiles and amphibians. Ray broke away from these concepts by defining fish as aquatic animals with no legs, gills, smooth skin or fins, and then dividing and subdividing them according to anatomical criteria (viviparous/oviparous, etc.) down to the species level[4]. The same

4 However, it is interesting to note that, perhaps fearing that this would move away from the prevailing classification, Ray later reintroduced cetaceans into the fish group.

applies to quadrupeds and birds. For insects, the criteria are based on their metamorphosis, the importance of which Ray fully understood.

5.4.5. *Observation of animal behavior*

Before moving on to the next century, let us quickly clarify what is meant by the observation of animal behavior. Of course, humans have always observed animals, regardless of material interest, religious belief or curiosity. Aristotle already showed an interest in this field and drew particular attention to certain species, such as bees. Later, there were also numerous observations and even treatises on behavior. In the Middle Ages, for example, the works on falconry by Frederick II of Hohenstaufen and of Gaston III of Foix-Béarn, known as Fébus, on hunting are surprising in terms of their accuracy and precision. The works of Hildegard of Bingen, Albert the Great and Bartholomew the Englishman are also worth mentioning. However, with a few exceptions, these writings often contain references to popular beliefs, ancient stories and travel accounts, as well as religious and symbolic elements. Furthermore, observations remain scattered, generally linked to certain species and, for practical reasons, to areas of human activity (medicine and pharmacology, hunting and fishing, falconry and animal training, beekeeping and various types of animal husbandry, including silkworm breeding, etc.). However, these various occurrences can, to varying degrees, be linked to a certain pattern that can be traced throughout the history of philosophy, with animals playing an important role, particularly in late antiquity, in the lively debates between Stoics, Neoplatonists, Skeptics and Christians. The arguments and often remarkable examples of performance, supporting intelligence or a lack thereof in animals and determining its nature and distance from human thought, thus paved the way for thinkers and theologians from the Middle Ages to the Renaissance. As a result, with a few exceptions, first-hand or reported observations of animals are rarely entirely objective; although they are frequently cited and given prominence, they are often exploited for controversial purposes, undermining their significance. Thus, in Montaigne, the sometimes excessive acknowledgment given to implausible examples, which he uses as skeptical arguments, contrasts with others that are drawn from personal observation[5].

In the 17th century, Descartes' theory of the "animal machine" caused a shockwave that reshaped the field of discussion through the debate that ensued on the soul of animals, which began in earnest after his death in the last third of the

5 For example: "When I play with my cat, who knows if she spends more time on me than I do on her?" *Essais*, Book II, Chapter 12.

century (with essentially theological implications) and continued into the 18th century (with issues that had become clearly anthropological). Admittedly, this thesis, which was not dogmatic in nature for Descartes, considered a radical ontological reductive view of the animal (the *beast*, to use the term of the time), albeit hypothetically, reduced to its body, which was itself reduced to a machine. However, supported by an argument based on observed behavior (the main criterion being the absence of proper language, which is distinct from simple communication), it was part of an approach that was already ethological in nature and, as such, encouraged future observations that would disprove it. This being said, among his successors, who were in a sense more royalist than the king (particularly Malebranche), the thesis became dogmatic and, leading to controversy, the example of animals was more often used as an argument to verify and reinforce doctrinal positions than as a genuine attempt to understand their behavior. It is therefore less the observation that counts than the demonstrative or argumentative approaches that instrumentalize it.

5.5. The 18th century: classification, observation, the question of the human species

Three different aspects of the approach to the animal world contributed to making the 18th century a pivotal moment that laid the foundation for the emergence of a true science of animals in the 19th century. On the one hand, there was Linnaeus' classification system, which finally provided a scientifically satisfactory framework for the study of animals and its development. On the other hand, in line with the previous century, there were advances in comparative anatomy and physiology. However, this in turn called into question the classification itself, raising new questions about the place of humans in the animal world and, ultimately, in connection with the history of philosophy, led to the beginning of a more objective and open observation of animal behavior.

5.5.1. *A detour into botany*

Faced with zoology's difficulties in resolving classification issues, a model was proposed: that of botany (Daudin 1983). For a long time, the science of plants had been seen as the "big sister" of zoology. Clearly, ahead of its time, although still heavily burdened by utilitarian and medical concerns, botany had made considerable progress since the Middle Ages, particularly in terms of classification (which was, in fact, often already applied to zoology). However, it faced a delicate dilemma: the choice between an "artificial method" and a "natural method". In their search for principles of classification, some favored this or that characteristic or set of characteristics deemed most relevant, whereas others declared it necessary not to favor any particular ones but to consider them all in each individual case, which greatly complicated the task but brought together groups, known as natural groups, that had always been perceived by common intuition, while the other approach drove them apart. Thus, through this dispute over characteristics, two completely different conceptions clashed: some sought a so-called "natural" method that would allow organisms to be classified in a coherent and satisfactory manner according to their obvious affinities, an undertaking that was very difficult and whose goal was constantly being pushed back; others, who were much more pragmatic, opposed them with a focus on efficiency above all else. For the latter, the main thing was to develop a convenient identification system (the "systematic" approach) that would allow even beginners to quickly and easily identify any plant encountered and add it to the general scientific database, using a minimal number of immediately visible characteristics, ideally just one. This method could be quite effective, but it was rightly called artificial, as it was not based on an effective understanding of the structure of organisms and often resulted in heterogeneous groups. The conflict of purposes was obvious: whether to truly understand or simply recognize, the objectives seemed difficult to reconcile. In addition, the problem was further complicated by a language issue: How should a unified nomenclature that reflected the classification system used be clearly developed? The names given to organisms varied considerably, not only from one country to another but also according to location, intended use (food, medicine, etc.), the treatments they underwent and sometimes even the time of year. Thus, while it boasted a certain superiority, botany was far from offering a united front or providing a simple model to follow. Nevertheless, it was indeed on the basis of this work, starting with Tournefort and Ray in the previous century and now, above all, with Linnaeus, that a sufficiently comprehensive classification of animals was gradually developed.

5.5.2. *Linnaean resolution*

Throughout his work, Carl Linnaeus (1707–1778) followed a simple and fundamental idea: God created an orderly world, assigning a place to every being, and seeking this order means, on the one hand, inevitably finding it, because it is inherent in the nature of things, and, on the other hand, serving God and drawing closer to him. Linnaeus was a deeply religious man: he was Swedish, raised in a very conservative Lutheran Protestant family by a father who was a vicar, then a pastor and a mother who was the daughter of a pastor, and was therefore not a man of doubt. Armed with these ideas and passionate about botany from an early age, he devised a new classification system at a very young age. Although it was simple and effective, based on the sexual characteristics of plants, it was not accessible to amateurs. He published it in 1735 in a work that was reprinted many times, each time updated: *Systema Naturae*. In these successive editions, the animal world was gradually integrated, and in the 10th edition (Linnaeus 1758), the binomial nomenclature (already partially in use), established in 1753 for plants, was generalized for animals. This nomenclature, which advantageously replaced the long Latin phrases that were in use at the time, consists of designating each species with a generic term followed by a specific term (e.g. *Panthera leo* for the lion). From then on, any new organism could immediately be given a place in the system that placed it in relation to all others. A true systematic and universal language was thus made available to the international scientific community, enabling each discovery to be classified and work to be carried out and exchanged within a common framework. Although it was immediately successful, Linnaeus also quickly encountered fierce opposition, particularly in France, owing to the largely artificial nature of his system, in which only certain elements – even if they were physiologically essential and not based on mere convenience – were retained at the expense of others. While these criticisms were directed particularly at his botanical concepts, they also extended to zoology.

5.5.3. *Resistance to the Linnaean system: Adanson, Réaumur, Buffon*

Opposition to the Linnaean system can be grouped together according to several types of reasons. The particularly controversial criticism of Michel Adanson (1727–1806), author of a *Natural History of Senegal*, written on the basis of his stay there and with only the first volume (Adanson 1757) devoted to shellfish being published, is less theoretically grounded and specifically targets nomenclature, arguing against Linnaean standardization, which he considers dictatorial, and calling for complete freedom of naming, including even terms from local dialects. As for Réaumur (1683–1757), he saw himself primarily as an observer and was rather indifferent to any classification system. He thus seems to embody the "usual divorce between the observer and the classifier" referred to by Daudin (1983, p. 148). Completely absorbed in his tireless work of description, he believed that the seriousness of science lay not in classifying beings but in studying their individual and collective behavior, particularly that of creatures whose delicate nature manifested the glory of the Creator more than anything. He devoted his *Mémoires pour servir à l'Histoire des insectes* (1734–1742), a treatise on ethology more than anatomy, to these creatures.

On a skeptical nominalist basis, Buffon (1707–1788), whose criticism did not exclude Réaumur (mocked for his wonder-filled finalism toward beings that were deemed insignificant), strongly criticized Linnaeus for segmenting a nature that was both continuous and made up of individuals, thereby falling into artificiality and arbitrariness. Even though he acknowledged the reality of species (of which he provided a precise definition as the interfertility of individuals), Buffon refused to further extend the categories he perceived as abstractions with no basis in nature, where everything is singular. However, his deepest critical motivation lies at another level: that of a major difference in perspective. As we have seen, Linnaeus' perspective, which is deeply religious, is theocentric, and his classification, which is supposed to reflect divine order, embodies an ideal guarantee of objectivity in this principle, whereas Buffon's perspective is, on the contrary, expressly anthropocentric (his monographs are also marked by strong anthropomorphism) and relativistic. Consequently, as Buffon did not believe in the objectivity of any classification, the approach taken in *Histoire naturelle* was to examine species in successive circles, from humans and their interests, beginning with domestic animals, then continuing with wild animals, first the most "noble" (the lion, etc.) and so on. Buffon also made a very interesting observation: if, beyond the narrow circle of the species, we accept the idea of groups formed on the basis of real affinities, then this implies a common origin for each animal family, a stock from which certain forms have been preserved and others have gradually degenerated. This was, in a way, turning Linnaeus against himself, undermining the idea of the fixity of species and proposing, albeit within narrow limits (these specific changes being confined to families), a transformist perspective (supported by others, including Diderot and, above all, Maupertuis). Lamarck (1744–1829) would not fail to remember this and give this idea a new scope. The position of the Jussieu family, Bernard (1699–1777), and his nephew, Antoine-Laurent (1748–1836), seemingly appears to be the most constructive. Rather than rejecting the Linnaean system, they sought an alternative natural method that would ultimately prove satisfactory. The result of a joint effort, the method developed by Antoine-Laurent de Jussieu (Genera Plantarum 1789), implemented the principle of trait subordination, which was later established by Cuvier. However, it failed to prevail over the convenience of the Linnaean system.

5.5.4. *The classification dilemma: unclassifiable intermediaries (pongo) and the place of man*

In the 18th century, discoveries and debates about intermediate beings increased: for example, the nature of corals (which Jean-André Peyssonnel, 1694–1759, established as animals), freshwater hydra (discovered and published in 1744 by Abraham Trembley, 1710–1784) and, in terms of gender distinction, the parthenogenesis of aphids (observed in 1740 by Charles Bonnet, 1720–1793) were questioned. This confusion in classification was not necessarily a problem, insofar as it reinforced the attachment to ambiguous beings valued by the chain of beings and thus the respective place of one another.

However, the problem took on a new aspect in the 10th edition of *Systema naturae* (Linnaeus 1758), in which Linnaeus subdivided the genus *Homo* into two types: *Homo sapiens*, or *diurnus*, comprising the American, European, Asian and African varieties, and *Homo ferus* (the savage man), on the one hand, and *Homo nocturnus* or *sylvestris* (the orangutan), on the other hand. Furthermore, in 1746, in *Fauna suecica*, Linnaeus admitted that his principles did not allow him to distinguish between humans and apes. This marked a turning point in how humans viewed themselves since, despite Linnaeus' assertion of the spirituality of the human soul, man, as a zoological genus, came to be included in the scholarly classification of animals in close proximity to great apes. The question of man's place was certainly not new and had given rise to a "classification dilemma" since the end of the 17th century, mainly owing to work in anatomy and physiology. The dissection of a chimpanzee by Tyson (1651–1708) in 1699 had been particularly striking. Tyson, assisted by Cowper, reported remarkable structural similarity with humans. The blurry boundary between humans and animals, already introduced in philosophy by empiricism and materialism (whose neo-Epicurean tradition was being reinforced by these new sources), revealed a disturbing scientific revelation.

Direct and indirect observations point in the same direction. This is also true for great apes, whose different species – chimpanzees, orangutans, gorillas (only identified in the mid-19th century) and bonobos (only identified in the 20th century) – remained largely undifferentiated at the time. According to a famous anecdote reported by Diderot in *Le rêve de d'Alembert*, Cardinal de Polignac is said to have exclaimed in front of the orangutan's cage in the Jardin du Roi: "Speak, and I will baptize you". Literature on this subject is increasingly abundant and well documented. On this basis, in note 10 of *Discours sur l'origine et les fondements de l'inégalité parmi les hommes* (Rousseau 1755), Rousseau asks the following: Are these Pongos, supposedly animals, not really wild men?

Rousseau's purely hypothetical and critical approach does not go any further, and Tyson's work was later corrected by Camper, Daubenton, Blumenbach, Goethe and Cuvier, who highlighted differences in the skeletons that enable humans to stand upright and hold their heads vertically. However, whatever is the case, these uncertainties in classification are not accidental; rather, they are also related to profound changes in the concepts that govern the observation of animal behavior, which is linked to a new theory of mind and knowledge.

5.5.5. Renewed observation and (partial) rehabilitation of animal capabilities

As we have seen, writings prior to the 18th century relating to observations – whether direct or merely reported – of animal behavior, even when they do not participate in controversial exploitation in the skeptical tradition, are not systematic in nature and, at best, are not based on anything other than purely empirical evidence. We have also seen how observation, regardless of how simple and transparent it may seem, is rarely detached from more or less implicit philosophical assumptions that often undermine its objectivity. Thus, the classification project had to contend with the metaphysical and theological idea of the chain of beings (which Voltaire ironically referred to in his *Dictionnaire de philosophie*, under the term "Chain of created beings"). However, in the 18th century and from the end of the 17th century onward, a shift took place in the way humans viewed animals. Among other reasons, this change was linked to a philosophical transformation in the understanding of knowledge and the formation of the mind. The latter, in connection with the critique of Cartesian theory and innate ideas, led to the foundation and rise of empiricism, which was renewed by Locke at the end of the 17th century. Indeed, in the 16th century, thinkers such as Montaigne (especially in *Apologie de Raymond Sebond*) and, in the 17th century, Gassendi (1592–1655), Marin Cureau de la Chambre (1594–1669) and La Fontaine (in his *Fables*, but even more explicitly in his *Discours à Madame de La Sablière*) developed a different discourse with regard to animals, but it was Locke who gave it the most solid and clear philosophical basis. With the English philosopher, whose ideas quickly spread across the continent, particularly in France, sensory experience was the foundation of all knowledge, which necessarily led to the unity of origin of the mind, whether it be animal or human. This idea was confirmed in the radicalized version of Lockean empiricism proposed by Condillac – from which he explicitly drew conclusions about animals in his 1755 work *Traité des animaux* (Guichet 2004) – and by the materialists, a leading philosophical force during the Enlightenment. Animal behavior thus needed to be reconsidered on entirely different grounds than those that had previously prevailed. Without any preconceptions about its capacities and limitations, each animal needed to be observed on its own terms without projecting a preconceived pattern onto its behavior.

It was Charles Georges Leroy (1723–1789), a friend of philosophers and well versed in their theories, who wrote articles for the *Encyclopédie* and drew concrete conclusions about the implications of their ideas in the field. As an experienced hunter, he developed the results of his observations in his *Lettres philosophiques sur la perfectibilité et l'intelligence des animaux* (published from 1762). In line with the writings of La Mettrie, Diderot, Helvétius and Condillac, he observed that intelligence in animals is proportional to their needs, as determined by their physical organization, their way of life and diet, their relationships with others, and the experience they accumulate as a result. Thus, quite logically, herbivores, which require little effort to obtain their food, will generally develop less intelligence than predators. However, we must also consider a strong individual factor; for example, an old, experienced deer being pursued by hunters will cross paths with a young deer to lead the hunters to follow the latter.

Furthermore, with this philosophical foundation, observing animal behavior makes perfect sense in relation to humans: by virtue of this common ground in sensory experience, animals become indicators of what it is that is truly innate in humans. Thus, Rousseau, again in the second *Discours*, bases his thesis on the observation of animals, arguing that self-love, as well as pity, is natural to humans (Guichet 2006, pp. 139–145, 276–282). These advances were definitely not achieved without difficulty. Descriptions frequently remain biased by anthropomorphism or anthropocentrism. The former is exemplified by Buffon in his *Histoire naturelle* (1749–1789), whose monographs are riddled with moralizing stereotypes about animals, often coupled with binary oppositions; thus, the lion, in a tradition dating back to Antiquity, is a figure of nobility opposed by the tiger, as cruel and deceitful as the lion is magnanimous and courageous; the same applies to the dog and the wolf, etc.

There is also a strong theocentric bias, especially with the development of natural theology, a movement that sought to find divine providence through creatures. Many scholars were associated with this movement to varying degrees, such as Bonnet and Réaumur. This was particularly true of entomology, as the behavior of insects, especially social insects, could easily be interpreted in a finalistic way. This did not detract from the scientific value of the work: such was the case, for example, with Réaumur's remarkable studies on the social organization of bees, observed in their beehive through a cross-section of glass.

5.6. Conclusion

From the Renaissance to the Enlightenment, as we have seen, the observation and classification of animals were far from being two clearly and closely linked movements, unlike what happened in botany (Daudin 1983, p. 49).

The first trend gradually developed knowledge of the anatomy, physiology and "habits" of animals, whereas the second, which was more abstract and focused on ancient references, applied theoretical thinking to the entire "animal kingdom", perfecting the methods and ways of organizing it to facilitate both the practical work of field researchers and amateurs and the theoretical understanding of scholars. Despite this disconnection, it is clear that it was the contribution of observation, mainly through advances in anatomy and the development of the microscope, that enabled animal classification to take on a scientifically stable and consistent form.

Thus, these two relatively independent movements continued to intersect until they converged in the 18th century due to Linnaeus' classification. At the same time, and in the same century, the observation of animal behavior gave rise to the first signs of scientific ethology.

5.7. References

Belon, P. (1551). *L'histoire naturelle des estranges poissons marins, avec la vraie peincture et description du Daulphin, et de plusieurs autres de son espèce, observée par Pierre Belon du Mans*. R. Chaudière, Paris.

Belon, P. (1555a). *L'histoire de la nature des oyseaux, avec leurs descriptions et naïfs portraicts retirez du naturel, escrite en sept livres*. G. Corrozet, Paris.

Belon, P. (1555b). *La nature et diversité des poissons, avec leurs pourtraictz représentez au plus près du naturel*. Ch. Estienne, Paris.

Carus, J.V. (1880). *Histoire de la zoologie. Depuis l'antiquité jusqu'au XIX*[e] *siècle*, P.O. Hagenmuller (trans.). Baillière, Paris.

D'Hondt, J.-L. (2007). *Histoire de la zoologie*. Ellipses, Paris.

Daudin, H. (1927). *De Linné à Lamarck. Méthodes de la classification et idée de série en botanique et en zoologie (1740-1790)*. Félix Alcan, Paris.

Delaunay, P. (1962). *La zoologie au XVI*[e] *siècle*. Hermann, Paris.

Duris, P., Gohau, G. (1997). *Histoire des sciences de la vie*. Nathan, Paris.

Foucault, M. (1966). *Les mots et les choses*. Gallimard, Paris.

Guichet, J.-L. (2004). *Traité des animaux, Condillac*. Ellipses, Paris.

Guichet, J.-L. (2006). *Rousseau, l'animal et l'homme. L'animalité dans l'horizon anthropologique des Lumières*. Cerf, Paris.

Guichet, J.-L. (2011). *Problématiques animales. Théorie de la connaissance, anthropologie, éthique, droit*. PUF, Paris.

Guyénot, É. (1957). *Les sciences de la vie aux XVII*[e] *et XVIII*[e] *siècles. L'idée d'évolution*. Albin Michel, Paris.

Hoefer, F. (1873). *Histoire de la zoologie. Depuis les temps les plus reculés jusqu'à nos jours*. Hachette, Paris.

de Jussieu, A.-L. (1789). *Genera plantarum, secundum ordines naturales disposita juxta methodum in Horto Regio Parisiensi exaratam*. Th. Barrois, Paris.

Knight, D. (1981). *Ordering the World. A History of Classifying Man*. Burnett Books, London.

Linnaeus, C. (1758). *Systema naturæ per regna tria naturæ, secundum classes, ordines, genera, species, cum characteribus, differentiis, synonymis, locis* (*Système de la nature, en trois règnes de la Nature, divisés en classes, ordres, genres et espèces, avec les caractères, les différences, les synonymes et les localisations*), 10th edition, J. Haak, Leyde.

Lovejoy, A. (1936). *The Great Chain of Being. A Study of the History of an Idea*. Harvard University Press, Cambridge.

Malpighi, M. (1669). *Dissertatio epistolica de bombyce*. Apud Joannem Martyn & Jacobum Allestry, London.

Mornet D. (1911). *Les sciences de la nature en France au XVIIIe siècle. Un chapitre de l'histoire des idées*. Colin, Paris.

Perrault, Cl. (1680). De la méchanique des animaux. In *Essais de physique, ou Recueil de plusieurs traitez touchant les choses naturelles*, volume III. J.-B. Coignard, Paris.

Petit, G. and Théodoridès, J. (1962). *Histoire de la zoologie. Des origines à Linné*. Hermann, Paris.

Pinon, L. (1995). *Livres de zoologie de la Renaissance : une anthologie, 1450-1700*. Klincksieck, Paris.

Roger, J. (1963). *Les sciences de la vie dans la pensée française au XVIIIe siècle*. Armand Colin, Paris.

Rousseau, J.-J. (1755). *Discours sur l'origine et les fondements de l'inégalité parmi les hommes*. *Œuvres complètes*, Gallimard, Paris.

6

Observation and Classification of Living Beings: Plants

Olivier Perru
S2HEP, Université Claude Bernard Lyon 1, Lyon, France

6.1. Introduction

Observing and classifying plants, from the Renaissance to the 18th century, was the focus of several research programs, whose essential contributions over three centuries cannot be covered in a summary work or encyclopedia article. We therefore believe that it is essential to focus on a few moments in the discovery of the natural world, on the one hand, and logical and scientific thinking, on the other hand, which we consider key to understanding the main scientific advances in botany and plant systematics from the 16th to the 18th centuries.

Discussing developments in botany during the Renaissance therefore requires consideration of the revision and reinterpretation of ancient texts, as well as the renewal (or re-founding) of this discipline, with new explorations and descriptions of the plant world, as well as new methodological tools (logic and dialectics) and techniques (printing, dissemination of books). The result was frequently disorganized scholarly work, characteristic of humanist scholars at the turn of the 16th and 17th centuries. The century of Descartes and Louis XIV marked the beginning of a systematization of naturalist knowledge, in which, in our view, *Port-Royal Logic* played an often-underestimated role that would perhaps prove more significant in the following century. Exploratory voyages and the proliferation of descriptive works provided a considerable amount of material that needed to be

History of Life Sciences and Medicine 2,
coordinated by Céline Cherici.

organized, which was done at the end of the century with de Tournefort's *Elements of Botany* (de Tournefort 1694).

Although there had been a latent issue of classification since the 16th century, which became truly evident with Tournefort's work, it was only during the Age of Enlightenment that it sparked debate and required choices to be made. Although Linnaean classification seemed to prevail, it was nevertheless supplemented and sometimes transformed by naturalists who were keen to move beyond the artificial nature of the Linnaean criteria to better understand the living reality of plant species. The 19th century proposed new syntheses for plants, as it did for animals, taking the evolution of knowledge into account. Beyond the issue of classification, it was the species itself that would pose serious difficulties in terms of how it was approached and the concrete realities of its existence. This is still a problem in the modern era, particularly in botany.

6.2. Botany in the 16th century: description and conceptual framework

6.2.1. *Describing and illustrating plants*

The 16th century was also a period of rediscovery of ancient texts in botany, particularly those of Dioscorides by Matthiole (1572). However, after 1540, there was also a period of change in practices (preservation, description and use of plants). Herbariums and botanical gardens appeared; scholarly works sought to describe known species as comprehensively as possible; medical use remained the primary focus, but botany and its home, the *hortus botanicus*, became less dependent on medicine. Valérie Chansigaud writes the following on this subject:

> In the years 1530–1540, scholarly practices underwent profound changes: dried plants were preserved in herbariums, and the medicinal garden (*hortus medicus*) broke away from its links with medical teaching to become a botanical garden (*hortus botanicus*). Above all, numerous works revolutionized botanical publishing and illustration. (Chansigaud 2009, pp. 18–19)

Various works on botany appeared at this time. Notable examples include those by Leonhart Fuchs (1501–1566), Rembert Dodoens (1517–1585) and Matthiole (Pietro Andrea Mattioli, 1501–1577).

For Matthiole, the observation and description of plants focused on the search for their medicinal applications, but for him, as for Dodoens, botany was in itself a science of observation and discovery. Matthiole's epistle and preface to his *Commentaires de Dioscoride* (Comments by Dioscorides) remain very much focused on medicine, even if there is a clear openness to the morphological description and illustration of plants. As a man of the 16th century, Matthiole believed that God, having created nature and the living world, taught Adam about living beings and expected humans to develop this knowledge, particularly about medicinal plants.

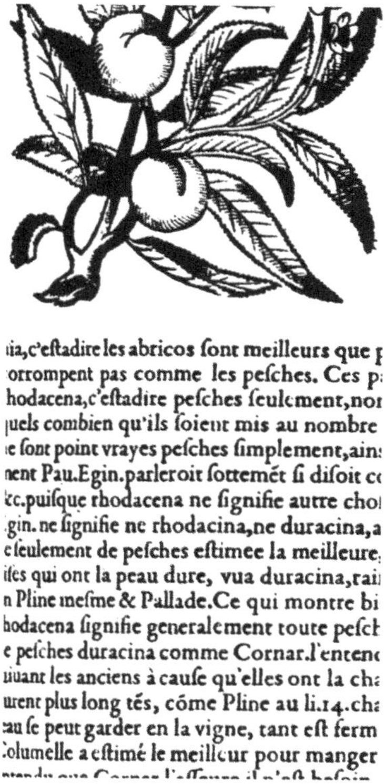

Figure 6.1. *The peach tree (Matthiole 1572, p. 159) (source: Google Books)*

Here is what he writes at the beginning of his epistle to Emperor Maximilian II, in the introduction to his work:

> And that God, in the beginning, infused the knowledge of all things into Adam, it is easy to speculate on the writings of Moses in Genesis, Chapter 1. For God gave power to our first fathers to be masters over the birds of the sky, the fish of the sea, and the other beasts of the

> earth, saying to them, "Behold, I give you every herb bearing seed which is on the face of all the earth, and every tree bearing fruit with seed in it; to you they shall be for food. And to every beast of the earth, and to every bird of the air, and to everything that moves on the earth, I give every green herb for food". From this, I say, we can conclude that God, having adorned man with such great gifts, did not forget to teach him the nature and virtues of all things by a single breath of the divine spirit. (Matthiole 1572)

Therefore, God could not forget to teach men "the aids and remedies that are plants, created solely for him" (Matthiole 1572, p. 9). Thus, men sought to learn more about plants and their medicinal properties. This is how Matthiole founded the science of apothecaries and botany. In the remainder of the book, the properties of simple plants follow the classification and principles of Hippocratic-Galenic medicine to the letter, with Matthiole simply adding various plant materials to Dioscorides' text, which were believed to be useful in treating various organ disorders.

Andrea Cesalpino (1524–1603) was the first naturalist of the 16th century who really cared about establishing a classification system for plants, such as an army in battle formation, according to a specific criterion: fruits. Linnaeus cited him in his *Fundamenta botanica* and, with little modesty, placed himself at the culmination of his project: "Cesalpinus is a fruit expert and the first true systematist; he classifies angiosperms and gymnosperms according to the number [of seeds]" (Linnaeus 1736, p. 6).

Cesalpino classified plants according to the shape of their flowers and, above all, according to the shape of their fruit and the number of seeds they contained. Among the 16th-century academics who wrote nothing on botany but gave institutional momentum to the discipline, Guillaume Rondelet (1507–1566) should also be mentioned. He is best known as a professor of medicine and zoologist, and author of the work *Histoire des poisons* (History of Fish). He actively contributed to the advancement of botany, zoology and medicine at the University of Montpellier, and his students included Jean Bauhin and Charles de l'Écluse.

6.2.2. *A reformed Aristotelian logic: definition following Pierre de La Ramée*

In the 16th century, academics described plants using genus, species and species subdivisions corresponding to their properties or relationships. They did not consider true classification, and most of them ignored the importance and functions of flowers, often preferring to rely on leaves or other characteristics to distinguish plants. In logic, genus and species are of Aristotelian origin; for Aristotle, genus (γένος) and species (εἶδος) are relative to one another, but only the species exists in reality. Unlike the modern understanding, in the natural world, the genus–species pair can be found at different levels: humans are a species of the genus animal, but a breed of dog can be considered a species of the dog genus (Aristotle 1883, p. 356). Philippe Lherminier writes the following regarding this:

> Genus is the similarity, generality, and extension that conveys unity; species analyzes ultimate difference and detailed understanding, determinable and empirical perception. In pure logic, genus is a universal concept that ignores taxonomic ranks. (Lherminier 2021, p. 49)

Genus and species are correlatives which, until the 17th century, were not considered from a classification perspective. According to Aristotle, in *Topics*:

> Genus is that which is essentially attributed to multiple things that differ specifically from one another. (Aristotle 1997, p. 12)

Genus is therefore a group of subjects that have something essential in common. Aristotle does not, however, systematically use the term "species" in logic to express how a subject differs from others. The French translation of *Topics* by Tricot describes the definition that "expresses the quiddity of the thing" (Aristotle 1997, p. 10). It would be better to say that it means the logos of what is: λόγος ὁ τὸ τί ἦν εἶναι σημαίνει. Tricot merely refers to "knowledge of quiddity" (Aristotle 1948, p. 256). In reality, contrary to popular belief, the emphasis on species, rather than the definition in Western logical tradition, mainly came from Porphyry. The reflection on the definition of "what is", of a particular nature, was taken up again during the Renaissance. Pierre de La Ramée wrote the following in 1555:

> A perfect definition is a definition composed of the constituent causes of essence, all of which are included in the genus and in the form: and thus, man is defined as a rational, animate being. For by genus (animate), we mean (as is dictated) a bodily essence full of life and meaning, which is what man is made of, and a part of the form, which if you add (reasonable), you will understand the whole form in its capacity for life, meaning, and reason. This form is also efficient, that is to say, it preserves the existence of the thing formed. (de La Ramée 1555, p. 58)

According to Pierre de La Ramée, it is therefore not simply a matter of describing all the accidents that occur to an individual and in the circumstances of their life; it is a matter of identifying what will bring about the essence of the species, a form that is essentially specific to that species and presupposes a genus to which it belongs and which is, in a sense, its matter. This seems to be a form of essentialism, where the specific form is understood both as the species defining a living being and as the cause that brings about and preserves "the being of the formed thing". This essence (in reference to Aristotle's to ti ên einaï – τὸ τί ἦν εἶναι – to be what one is) links the species and the individual, but we are far from the metaphysical nature of Aristotle's train of thought. We have moved from the being to the thing that is and to the thing itself. In reality, as in botany, there is never a single form that can be used to define an object; the specific form will be a relationship between forms. Philippe Selosse thus applies this way of defining objects in *Res herbaria*, taking an example from Caspar Bauhin's *Pinax Theatri Botanici* (Bauhin 1623) concerning gentian (Figure 6.2):

> If we take a particular genus, such as *Gentiana*, we see that all plant names can easily be placed in a logical division tree [...]. Three stages of logical division therefore occur in the genus *Gentiana*, as revealed by the nomenclature. The supreme genus *Gentiana* is subdivided into species, including *Gentiana alpina*; which then functions as a (lower) genus, insofar as it is subdivided into species, including *Gentiana alpina major*; and this functions as a (lower) genus, insofar as it is subdivided into indivisible species, including *Gentiana alpina major lutea* [...]. A plant nomenclature was therefore established during the Renaissance, but with a definition-based perspective of summarizing all levels of logical division (Selosse 2004). (Selosse 2008, pp. 68–69)

From the Renaissance to Descartes, definitions were therefore very diverse. Logical subdivision would characterize the perfect definition – it is orderly, descending, dynamic (the levels were starting points or end points of divisions), undefined, reticular (in the sense of a network of relationships between forms), etc. However, this is not a classification: these characteristics contribute to defining what a particular plant is. The perfect definition therefore presupposes the procedure of logical subdivision. Thus, plants (or other objects) can also be defined imperfectly, which is more concise and closer to description and illustration, thus expressing what a particular species of plant is.

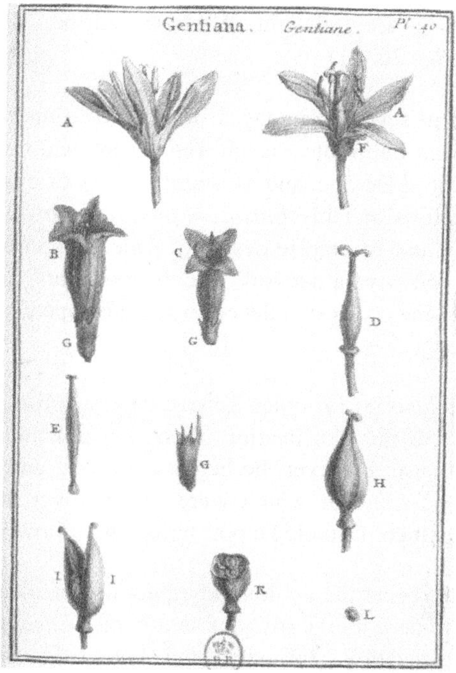

Figure 6.2. *Plate 40, gentian (de Tournefort 1694) (source: gallica.bnf.fr). For a color version of this figure, see www.iste.co.uk/cherici/history2.zip*

6.2.3. *The scholarly heirs of the Renaissance: the contributions of Charles de l'Écluse and Peiresc*

6.2.3.1. *Charles de l'Écluse*

Charles de l'Écluse (1526–1609, who adopted the Latin name Clusius) is known for his work in botany but also for his correspondence, for example, at the end of his life, with Peiresc. Charles de l'Écluse was clearly a student of Rondelet and a graduate of the Montpellier Faculty of Medicine. He was interested in exotic plants brought back from Asia and South America for medical use. He translated three Spanish works on these new medicinal resources, as well as the Latin *Herbarium* and *History of Plants* by Rembert Dodoens.

> Already known as a humanist when he arrived in Montpellier on October 3, 1551, it was with Rondelet's support that he developed a taste for botany and prepared himself, during a three-year stay in the south of France, for his fruitful explorations of the flora of the Iberian

> Peninsula and of Central and Western Europe. (Planchon and Planchon 1866, pp. 30–31)

Clusius was fluent in several languages, which led to his significant work as a translator, including from Latin into French. He was an avid traveler, particularly in Mediterranean and Central Europe, and as a result of his extensive herbarium work, he wrote botanical treatises, notably *Rariorum plantarum historia* (L'Écluse 1601). His work pioneered the use of precise descriptive methods in botany (Gérard 1871, p. 28). He was at the center of a network of correspondents, as was Peiresc a little later. In both cases, the correspondence was accompanied by exchanges of specimens and knowledge.

For Charles de l'Écluse, as for other Renaissance botanists, the description of a plant's qualities refers to the glorification of the Creator and often relates to the cultivated plant and its use; however, he began collecting and describing plants of different exotic natures, some of which were of unknown interest but which he intended to grow in a garden. Danièle Duport writes the following:

> Travel accounts reveal the apologetic aspect of botanical description. They invite us to taste all the fruits of nature, all its benefits, to expand the realm of sensation, curiosity, and knowledge in a renewed admiration of Creation. Like Rabelais' journeys, the products of foreign lands and their abundance of unusual plants only serve to praise unlimited diversity or even to deem something potentially monstrous as normal, or at least unusual, in the natural order of things. Everywhere, here and elsewhere, the traveler discovers that man is pursuing his task of cultivating God's extraordinarily varied garden and taking inspiration from the providential complexity of nature. (Duport 2001, p. 212)

Around 1604–1605, Peiresc, who was still young and living in Montpellier to study law, sent Charles de l'Écluse in Leyden seeds from various plants and bulbs, as well as an inventory of his herbarium in Languedoc. On February 25, 1604, he wrote the following:

> Two or three days ago, I noticed in your book that you have never seen the Pancratium flower that grows near Narbonne, which made me greatly regret not having the opportunity to send you some bulbs from Montpellier, where Monsieur Riché has a large quantity of plants that bloom every year around September and October, producing several white flowers shaped like extravagant bells surrounded by a few long, narrow leaves that make them very beautiful. Moreover, the

> scent is almost identical to that of an orange tree, albeit much sweeter and more pleasant, so much so that the ladies of Montpellier enjoy wearing bouquets made from orange tree leaves, which resemble branches of flowering orange trees. (Peiresc 1898, p. 947)

The book Peiresc refers to is *Rariorum plantarum historia*, published in 1601. In the same letter, he thanks his correspondent for sending him this book. Pancratium is a sea daffodil, which grows on the sandy coasts of the Mediterranean; thus, its location around Narbonne is plausible. "Monsieur Riché" is Pierre Richer de Belleval (1555–1632), founder of the Montpellier Botanical Garden. Letter 412 details the layout of the botanical garden. Bulb plants, such as Pancratium, were very popular at the time. Clusius lived in Holland, where tulip cultivation was booming until the tulip bubble burst in 1637. Finally, Peiresc's letters mention another plant, which he calls the Tragacantha seed. This was *Astragale de Marseille*, later defined in 1753 as *Astragalus tragacantha*. It is a *Fabaceae* forming a branched, thorny bush. In 1604, Peiresc requested that this plant and various other Provençal plant species be sent from Marseille to Clusius in Leyden. Finally, when the apothecaries of Marseille claimed that *Astragalus maritimus* did not produce seeds, Peiresc sent a box of *Astragalus* roots and various other plants to his correspondent in 1605 (Legré 1900, p. 410).

From 1601–1605, a series of seven letters from Peiresc to Clusius, as well as an inventory of plants, were published at the end of the 19th century by Tamizey de Laroque. Peiresc's correspondence, which we will now examine, plunges us into the core network of these botanical exchanges between the 16th and 17th centuries.

6.2.3.2. *Peiresc*

Nicolas-Claude Fabri de Peiresc (1580–1637) was a humanist scholar, renowned figure of the Renaissance, and a magistrate in Aix-en-Provence, whose scientific curiosity knew no bounds. He was interested in Byzantine manuscripts as well as astronomy, anatomy and plants. On his estate in Belgentier, near Toulon, he cultivated plants from the East, notably Persian jasmine, which he was said to have introduced around 1630. Peiresc was passionate about jasmine (Figure 6.3) and orange trees, and we know that he brought several plants to Belgentier and cultivated them, including Adam's apple lemon tree (mentioned in 1632), Indian yellow jasmine (1629), Mecca squash (1630) and cinnamon laurel (1632). Peiresc helped acclimatize many of these plants to France; he had a vast network of correspondents who exchanged various collectibles, books and even plants and animals, both living and dead. This is how he introduced the Angora cat (from Turkey) to France. His letters are full of anecdotes about sending and receiving seeds and plants, which sometimes arrived at their destination in poor condition.

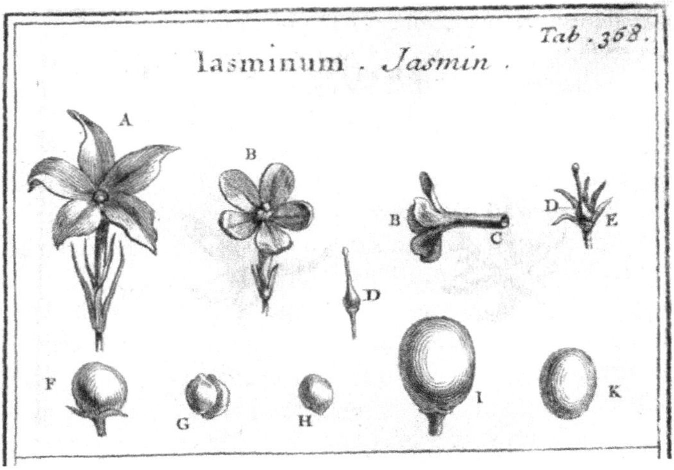

Figure 6.3. *Jasmine (flower and fruit) by de Tournefort (1700), table 368 (Latin edition of* Elements of Botany*) (source: Wikimedia*[1]*)*

The first traces of Peiresc's interest in botany are found in his correspondence with Charles de l'Écluse. In 1606, while traveling, he met him in Leiden, just as he met the physician and botanist Mathias de l'Obel (1538–1616) in England. After 1608, there were few traces of Peiresc's botanical activity until 1626, when his correspondence mentioned the plants he was growing in Belgentier and Aix. In terms of botany, Peiresc was mainly interested in designing gardens for private estates and acclimatizing various species; this is how he landscaped the gardens of the archbishopric in Aix on behalf of the archbishop at the time, Alphonse-Louis de Richelieu (1582–1653), who was brother to the cardinal minister.

> We had trouble finding clean abalone, and not only took some from the local area, but also from Merargues, St Cannat, and Jouques, for the moon and the season pressed us onward. During the crescent of the present moon, we will try to plant the buoys in the flowerbed, after having decorated the bottom of the gallery wall with lilacs and broom, for lack of any other suitable plan. We will sow laurel on the side facing north, and for the south side, we will see about planting jasmine, waiting to see if the Archbishop would like something else there. If these elm trees do not suit him because they have small leaves, we can bring in grafts from the large-leaved trees of Orléans and have them planted in a year or two. (Peiresc 1896, pp. 440–441)

1 See: www.flickr.com/photos/biodivlibrary/8161853233.

Further information about various plants appeared in Peiresc's correspondence in the late 1620s. From 1629–1632, Peiresc fled the plague that had struck Aix and settled in his property in Belgentier, where he seems to have devoted part of his time to his plantations and naturalist observations. On December 15, 1629, he wrote to Dupuy that he had discovered petrified leaves and shells, which were very different from Mediterranean specimens and rather exotic. However, Peiresc did not dwell on this and resumed his description of the plants of the Toulon region, notably Styrax:

> Only the shrubs and plants that grow in the most uncultivated areas of the hills surrounding us would be capable of providing pleasant enjoyment to those who do not mind seeing the mastic gum on our lentisk, and that of Styrax, which is fragrant and sweet, on a shrub that produces flowers like the orange tree, and leaves like those of the quince tree, whose species is only found here and within a mile or so of the surrounding area. (Peiresc 1890, pp. 208–209)

The lentisk, *Pistacia lentiscus*, is the mastic tree with a resinous scent; the *Styrax* genus includes approximately 100 species, with fragrant resin known as benzoin. *Styrax officinalis* is found on the shores of the Mediterranean. Peiresc's life and letters contain numerous references and descriptions of exotic plants, as well as their cultivation during the years he spent in Belgentier. However, this somewhat disordered interest is limited to a collection of names and descriptions corresponding to plants grown in Mediterranean gardens. The main aim seems to have been to establish a garden (or forest) for the use of plants or trees (for decorative, food or medicinal purposes or for wood). The emphasis on the garden was a sign of its development in the classical age. Peiresc shows no classification or even understanding of the functions of the various parts of plants. At the turn of the Renaissance and the classical age, botanists remained focused on the qualitative description and enumeration of the plants they encountered. However, according to Julius von Sachs:

> We are seeing an increase in comments on the identity or non-identity of species, on the differences that exist between plants that were previously thought to belong to the same family. de l'Écluse's descriptions are truly scientific in nature [...]. What is most surprising about these descriptions [...] is the way in which the authors in question neglected to study the flowers and fruits. (von Sachs 1892, p. 19)

Even de Tournefort, who based his classification on flowers at the end of the 17th century, did not fully understand their reproductive function. It was not until Linnaeus that classification was organized around plant reproduction.

6.3. Botany in the 17th century

6.3.1. *A discipline becoming institutionalized: the King's Garden*

In 1626, Louis XIII decided to purchase land for growing medicinal plants and teaching natural history and therefore botany. Guy de la Brosse (1586–1641) was the first superintendent of the royal botanical garden, commonly known as the "King's Garden". In the 17th century, this garden was a place of learning, and following La Brosse, it was marked by leading figures such as Guy-Crescent Fagon (1638–1718) and de Tournefort (1656–1708). It would be impossible to retrace the long and complex institutional history of the garden here until 1793, when it was transformed into the French National Museum of Natural History. Notably, particularly during the reigns of Louis XIV and Louis XV, this institution flourished both as a place where institutional science was developed and as a reference point for the teaching of chemistry and botany.

6.3.2. *A new logical tool for defining and classifying:* Port-Royal Logic

Despite the proliferation of botanical gardens and the creation of the King's Garden, progress in botany was slow in the 17th century. In this field, as in others, there was a need for a method of organizing the living world. The Cartesian movement and *Port-Royal Logic* proved to be invaluable in this regard. Was there a resurgence in logic after Descartes? In part, yes, because a new link was established between logic and mathematics, particularly with the successive editions of *Port-Royal Logic*, which continued until 1683. Father Lamy emphasized the role of logic in learning; Descartes himself never denied that scholastic logic was useful for young minds. After *Port-Royal Logic*, the desire to educate minds led to a synthesis between the logic inherited from scholasticism and the model of geometric reasoning promoted by Descartes. Malebranche and Lamy, like Arnauld and Nicole, shared this perspective. Referring his readers to *Port-Royal Logic*, Mabillon himself wrote the following:

> We must therefore use logic to accustom ourselves to thinking and reasoning correctly, to correcting common sense and forming judgments, and not to spoiling it, as happens to those who study it only to learn how to argue formally in public disputes, that is to say, to

> quibble. The *Art of Thinking* shows how logic should be used, and the two discourses at the beginning, concerning the ideas and rules to be observed in order to make a fair and impartial judgment on many things, should be read with particular attention. (Mabillon 1691, p. 335)

In reality, *Port-Royal Logic* would become part of the methodological approach of the 18th century, both in the intellectual culture of the time and in scientific progress, including botany. In 1788, de Caritat de Condorcet wrote the following:

> The Duke of Chevreuse [...] had been educated in the principles of the new philosophy of his time by the hermits of Port-Royal: for every century and almost every generation has its own, which, always slandered by contemporary foolishness, later becomes one of the prejudices of posterity. The work, long known under the title *The Art of Thinking*, was the result of these learned conversations. (de Caritat de Condorcet 1847, pp. 306–307)

The relationship between *Port-Royal Logic* and the subsequent development of science is apparent here. In Book IV of *Port-Royal Logic*, the science that the authors are seeking is Cartesian science, the science of "what can be known clearly and certainly" or "what we can hope to know" (Arnaud and Nicole 1683, p. 295). The guiding principle is to seek knowledge of objects that are proportionate to us.

> The greatest summary that can be found in the scope of science is never to apply oneself to the search for anything that is above us and that we cannot reasonably hope to understand. All questions concerning the power of God are of this kind. (Arnaud and Nicole 1683, p. 295)

The skeptical Jansenism of Arnauld and Nicole, by asserting the infinity and power of God as absolute and assigning them to the realm of faith, rejects metaphysics and even speculative theology, and limits human reason to verifiable questions, particularly in the sciences of measurement.

Can we further clarify the possibilities opened up by *Port-Royal Logic* for the development of a science such as botany? Undoubtedly, yes, but only if we affirm that it is part of a broader movement aimed at defining and naming species. According to Arnauld and Nicole, it is a matter of finding the difference between one species and another in philosophy, as in science. The name covers a definition expressing the most appropriate difference between species of a genus. This correspondence between name and definition (which contains the idea of difference) can be found in de Tournefort's *Elements of Botany* (de Tournefort 1694). The emphasis, in *Port-Royal Logic*, on the correspondence between the name and the definition, which is based on the genus and the essential difference, is echoed in the classification systems of the 17th and 18th centuries (Tournefort and Linnaeus). The sign can refer to something absent or something present (a sign implying the thing, or a certain presence of the thing), it establishes the link between reality and the idea, between what is known by the senses and what comes from the mind, it mediates through figures, numbers and words. A sign is, above all, a word: it synthesizes the definition and properties and designates the known thing.

> Although a thing in one state cannot be a sign of itself in that same state, since every sign requires a distinction between the thing represented and the thing that represents it; nevertheless, it is quite possible that a thing in a certain state may represent itself in another state, and that thus the distinction of state alone is sufficient between the thing representing and the thing represented: that is to say, the same thing may be in one state as something signifying, and in another state as a something signified. (Arnaud and Nicole 1683, p. 54)

Arnaud and Nicole give the example of the sign of bread in the Eucharist. The transposition of this theological figure into the secular realm could imply a materialization of the relationship between the signifier and the signified, the representative and the represented, in terms of the logic of signs. We can then understand the significance of the link between the definition of a plant or animal species, its scientific illustration and the biological individuals associated with it.

The relationship between *Port-Royal Logic* and science is complex.

> It is difficult to say outright that it paved the way for the development of natural history in the following century. However, as an emblematic work of knowledge in the classical age, it laid the foundations for the formation of complex systems of signs for thinking about the surrounding world. (Perru 2012, p. 415)

On this subject, in *The Order of Things*, Michel Foucault writes about signs in *Port-Royal Logic*, saying that they are responsible for representing and displaying the world, "juxtaposing it according to an indefinably open surface, and from it, pursuing the endless unfolding of substitutes that we can think of" (Foucault 1966, p. 75).

6.3.3. *Journeys of exploration and field observations: Father Charles Plumier in the West Indies (1646–1704)*

Charles Plumier was a member of the Order of Minims, a religious order during the reign of Louis XIV. He made three journeys to the West Indies on the King's orders; the second journey was a success: upon his return, Louis XIV appointed him as a royal botanist, and he wrote *Description des plantes de l'Amérique* (Description of Plants in America) (Plumier 1693). In this work, he was particularly interested in the ferns he discovered in the West Indies to the point that he devoted specific work to them known as *Traité des fougères de l'Amérique* (Treatise on Ferns in America), which was published after his death in 1705. With respect to ferns, the English traveler Martin Lister, who had met Plumier in Paris and appreciated his work describing approximately 600 plants, wrote the following:

> In his vast collection of plants, I noticed that vines and ferns were by far the most numerous. He had an incredible number of species of each of these genera. (Lister 1873, p. 77)

Plumier did not bring back any animal or plant specimens as everything was lost in a shipwreck; however, he kept the drawings he had made on site of the fauna (especially fish) and flora of the West Indies. He proved to be an excellent naturalist and descriptor of the species he encountered, as well as a very good artist. In the preface to his 1693 work, he recounted the circumstances of his recruitment for these scientific journeys, when he was living in Marseille and already had good knowledge of Mediterranean and Alpine flora. Plumier died of pleurisy in 1704, just as he was about to embark at Cadiz on a fourth voyage, with Peru as his destination. He is a good example of a traveling naturalist who increased knowledge in the 17th and 18th centuries without seeking to classify what he found; he collaborated with de Tournefort from 1678 onward.

In the preface (unpaginated) to his *Description des plantes de l'Amérique*, Plumier writes that a botanical demonstration given at the Trinité-des-Monts convent in Rome by Brother Philippe Sergeant, a pharmacist, led him to abandon his studies in mathematics in favor of botany. His superiors in the Order of Minims recalled him to Marseille and granted him permission to collect plants in the southern Alps and along the Mediterranean coast. It was at this time that Michel Bégon (1638–1710), intendant of the galleys in Marseille and to whom (according to de Tournefort) he dedicated the discovery of *begonia*, contacted him about the trip to the West Indies:

> To satisfy the King's orders, he wished to find someone who could make the journey to our West Indies (where he had been Intendant), so as to search for all the rarest and most curious things that nature produces there. He made the proposal to Mr. Surian, who was highly capable, not only in the knowledge of plants, but also in the secrets of chemistry; and simultaneously commissioned him to find someone who could assist him in carrying out this plan. Mr. Surian made the proposal to me, and I gladly accepted... (Plumier 1693)

The naturalist traveler method combines inventory with description:

> Like most naturalist travelers, he compiled inventories of fauna and flora, wrote descriptions, and named plants after great botanists of the Renaissance and the 17th century. He made hundreds of drawings himself, collected samples to create a herbarium, and took seeds to try to acclimatize these new species in France. (Hanomou 2008, p. 55)

Plumier proceeded as follows: his knowledge of botany enabled him to compile an inventory, which he described, named, and ultimately drew. However, his method was undoubtedly influenced by de Tournefort's work and the latter's creation of new genera. In the plates he drew, this method was not fully developed (Figure 6.4):

> These plates are not true botanical plates, in the sense that they do not show all parts of the specimen: roots, leaves, inflorescences, fruits, or seeds seen from different angles. Plumier very rarely drew complete plates and generally only showed details of the floral parts: stamens, pistils, corollas, etc.; or of the fruits. Could this be the influence of his friend Tournefort's classification system, which led him to only present the parts relevant to the system? We do not know, but we can make a guess, however, on the influence of the engravings in *Elements of Botany* in the series of very detailed plates he produced, which were published in the work entitled *Nova plantarum americanum genera* in 1703. [...] Plumier's drawings are, of course, important for the knowledge and recognition of the flora of the New World and the establishment of new genera, such as those established by Tournefort, for example. (Hanomou 2008, pp. 61–62)

Figure 6.4. *Plates XXVI and XXIX (genus* Lonchitis*) (Plumier 1693) (source: gallica.bnf.fr). For a color version of this figure, see www.iste.co.uk/cherici/history2.zip*

6.3.4. Elements of Botany *(1694) by Joseph Pitton de Tournefort (1656–1708): moving toward a classification?*

De Tournefort was the great botanist of the late reign of Louis XIV. Born in 1656 in Aix, he was interested in medicine and botany, traveled extensively, studied under Magnol in Montpellier, and eventually became a lecturer at the King's Garden in Paris, where he was protected by Fagon. Today, he has paved the way for the recognition of floral morphology and, to a lesser extent, the sexual characteristics of plants, with a view to classification. De Tournefort explained his method as follows:

> The method we have followed is usually based on the structure of flowers and fruits. One cannot deviate from it without encountering strange difficulties, as was shown in the first part. However, we acknowledge that this method is not universal, and we have not

> hesitated to use other parts of plants when we deemed it necessary. (de Tournefort 1694, warning)

Here, we clearly look at the specific structure of the flower and fruit, especially the shape of the corolla, but not the dynamics of the plant's reproductive system. Indeed, he was completely unaware of the role of the stamens and pistil in fertilization, although he did take them into account, morphologically speaking, in his description of genera:

> For Tournefort, stamens were merely drainage channels and pollen was excrement. (Hoquet 2007, p. 55)

Eleven years before *Elements of Botany*, the 1683 edition of *Port-Royal Logic* argued for correspondence between names and definitions on the basis of genus and essential difference. This is precisely what de Tournefort proposed. As a staunch Cartesian, if we are to believe Le Bovier de Fontenelle (1740, p. 170), de Tournefort "contributed to the transition from the logical notion to the naturalistic notion" of genus (Drouin 2008, p. 24). It is necessary to combine "the name of each plant" with the idea of the trait that "essentially distinguishes plants" (de Tournefort 1694, pp. 1–2). From there, we can trace the latter back to the genus, and only approximately 600 plant genera names should be retained in the study. However, the emphasis on genera does not simplify the approach to species: the nomenclature is still the same as that inherited from the Renaissance, with several qualifiers for each species. The genus Veronica is very well described by a text accompanied by a plate (plate 60; Figure 6.5):

> Veronica is a genus of plants whose flower A is a rosette with four lobes, in the center of which there is a hole B, through which the pistil D comes up from the middle of the calyx C. When this flower has faded, this pistil becomes a fruit EG divided into two capsules F, H, in which there are some very small seeds K in some species, and rather large ones in others, which have a fairly significant indentation. (de Tournefort 1694, p. 120)

The description is therefore fairly accurate, but the nomenclature retains some inconvenient subdivisions dating back to the Renaissance: *Veronica mas supina et vulgatissima*, *Veronica spicata latifolia*, *Veronica spicata longifolia*, *Veronica spicata minor*, *Veronica pratensis serpillifolia*, *Veronica Alpina frutescens*, etc. These names became obsolete with the introduction of Linnaeus' binomial nomenclature.

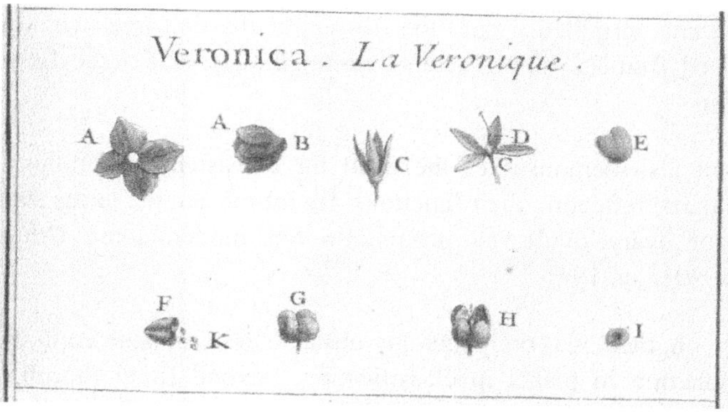

Figure 6.5. *Plate 60, Veronica (de Tournefort 1694) (source: gallica.bnf.fr). For a color version of this figure, see www.iste.co.uk/cherici/history2.zip*

From the genera, which truly appear to be the core of botanical knowledge, de Tournefort created classes on the basis of the shape of flowers (corolla) and inflorescences; he clearly distinguished between herbaceous plants and shrubs and trees, which Buffon praised him for. However, de Tournefort's classification of genera and classes remained incomplete and controversial, as it is dependent on a complex nomenclature. Even during de Tournefort's lifetime, as early as 1694, Camerarius (1655–1721) demonstrated the existence of plant sexuality. He distinguished between hermaphroditic, monoecious and dioecious flowers. He was one of the botanists of his time who was not satisfied with just describing, but who also experimented:

> He noticed that, after removing all male mulberry trees from a female mulberry tree, the latter bore fruit without embryos; in mercury, after removing the male flowers, he observed that the female flowers produced seeds, none of which were fertile. (Magnin-Gonze 2015, p. 180)

After de Tournefort's death, Hermann Boerhaave accepted and taught about the existence of sex in plants at the University of Leiden. In France, Sébastien Vaillant (1669–1722), a student of de Tournefort, introduced and explained the sexual function of flowers in the King's Garden in 1717.

> It was a pistachio tree from the King's Garden that finally confirmed the demonstration. After observing a tree of this species in the Apothecary's Garden, which flowered every year but did not bear fruit, Vaillant took pollen from another tree growing in the seedling

> garden and sprinkled it onto the flowers of the first tree, which then produced fruit. (Synowiecki and Noûs 2020, p. 455) (see Bertrand (2018)).
>
> Vaillant also demonstrates the need for consistent terminology for floral parts, reflecting their functions. He introduced the terms stamen, filament, ovary, ovule, and placenta in their modern sense. (Magnin-Gonze 2015, p. 194)

From then on, there was no longer any obstacle to be considered in the criteria of sexual reproduction in plants in classification, beyond floral morphology alone. Linnaeus followed from the work of de Tournefort and Vaillant but also brought about a substantial transformation in their approach to classification (Vaillant 1718).

6.4. The epistemic revolution of the 18th century

6.4.1. *The Linnaean classification*

Carl Linnaeus (1707–1778) had a stroke of genius in *Fundamenta botanica* (Linnaeus 1736), a very short 36-page work in which he very specifically and comprehensively introduced the sexuality of plants as a criterion for classification. After explaining that de Tournefort was a corollist and that Pierre Magnol (1638–1715) was a calicist associated with the fructists, he wrote the following:

> I, myself, am a sexualist, according to the number, proportion, and position of the stamens in relation to the pistil. (Linnaeus 1736, p. 6)

This is precisely what makes Linnaeus so original: he associated the stamens and pistils as male and female organs of sexual plant reproduction and used them as a classification criterion. Linnaeus' "I, myself" emphasizes his creativity and the relative novelty of his system. This was the great botanical revolution of the 18th century. We know Buffon's criticism of this classification, which he considered to be artificial: it led to placing plants, as morphologically diverse as roses and strawberries, in the same family. However, Linnaeus' point of view was anatomical and physiological rather than morphological. He pointed out that every living being comes from an egg and that this is therefore also the case for plants. In Linnaeus' text, the word *egg* should actually be read as "ovule".

> The cotyledons come from the vitellus of the egg, in which the living core is rooted. Therefore, the seminal leaves that surrounded the hilum are the same thing. It is thus proven that the descendant does not

solely come from the egg or from the seed but that its generation comes from both simultaneously. (Linnaeus 1736, p. 16)

Linnaeus therefore refers to the fruit, but he considers that fruiting over time includes the entire flower and fruit. He details the composition of the flower and the reproductive organs it contains (*genitalia*): the anthers are the male reproductive organs, and their dust (pollen, whose reproductive behavior would only be fully understood by Koelreuter (1733–1806) and Sprengel (1750–1816)) is a true seed; the stigmas are the female *genitalia*. The fall of the seed from the anthers onto the stigmas determines fertilization. We can see how this precise description is reminiscent of the analogy of fertilization in animals and even in humans. Linnaeus then specifies the existence of flowering plants that are either male or female (in other words, dioecious species), hermaphroditic plants (today, we would specify that these are flowers with both traits) or androgynous plants (having both types of flowers, which should be specified as being in different places on the same plant, for example, on the same stem). We can see that Linnaeus does not develop his subject very much, merely naming the various possible cases.

The same would apply to classification, according to Thierry Hoquet:

> Adanson also notes that the Linnaean system is "artificial": its classes proceed according to simple divisions and mechanically decline *Monoandries*, *Diandries*, etc., giving precedence to a logical principle rather than to the perception of natural groupings. (Hoquet 2008, p. 58)

Linnaeus was therefore criticized for the mechanical and artificial nature of his groupings, particularly on the basis of the number and form of the sexual organs of plants. Thus, the automatic subdivisions of Linnaeus' classification were no longer able to account for the proximity and diversity of natural groups. This became the great debate of the 18th century on the priority of a classification method: should it be natural or artificial? Before entering into this debate, we must also consider the effectiveness of Linnaeus' classification system:

> Thus, because the sexual system is based on a small number of principles, it leaves no room for ambiguity, making it easier to identify a plant, even if it does not have the extreme simplicity that has been attributed to it. (Hoquet 2008, p. 59)

Figure 6.6 highlights the nature of the various groupings established by Linnaeus, depending on whether the stamens are free or adhere to one another or to the pistil. This method is therefore the basis of Linnaeus' classification.

Furthermore, because he was rather logical, Linnaeus undoubtedly considered the application of the genus and species predicables in botany; he may have considered difference to be the constituting essence and definition of the species. In any case, he should be praised for his approach to the species through the genus name and the species name, and possibly the common name, in his classification of the living world (Raynal-Roques and Roguenant 2008).

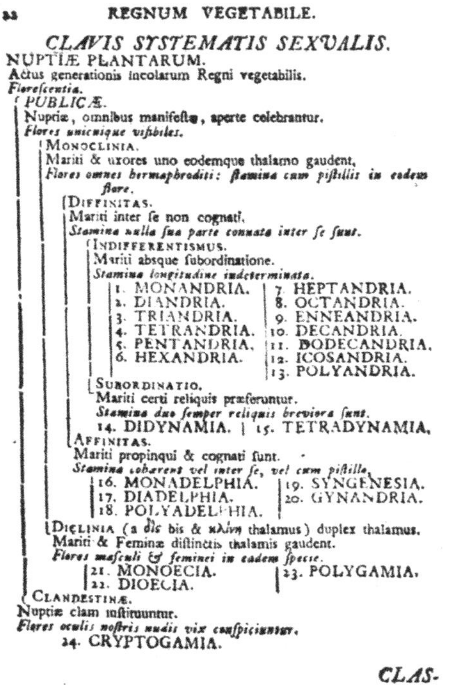
22 REGNUM VEGETABILE.

CLAVIS SYSTEMATIS SEXUALIS.

NUPTIÆ PLANTARUM.
Actus generationis incolarum Regni vegetabilis.
Florescentia.

PUBLICÆ.
Nuptiæ, omnibus manifestæ, aperte celebrantur.
Flores unicuique visibiles.

MONOCLINIA.
Mariti & uxores uno eodemque thalamo gaudent,
Flores omnes hermaphroditi: stamina cum pistillis in eodem flore.

DIFFINITAS.
Mariti inter se non cognati,
Stamina nulla sua parte connata inter se sunt.

INDIFFERENTISMUS.
Mariti absque subordinatione.
Stamina longitudine indeterminata.

1. MONANDRIA. 7. HEPTANDRIA.
2. DIANDRIA. 8. OCTANDRIA.
3. TRIANDRIA. 9. ENNEANDRIA.
4. TETRANDRIA. 10. DECANDRIA.
5. PENTANDRIA. 11. DODECANDRIA.
6. HEXANDRIA. 12. ICOSANDRIA.
13. POLYANDRIA.

SUBORDINATIO.
Mariti certi reliquis præferuntur.
Stamina duo semper reliquis breviora sunt.
14. DIDYNAMIA. | 15. TETRADYNAMIA.

AFFINITAS.
Mariti propinqui & cognati sunt.
Stamina cohærent vel inter se, vel cum pistillo.
16. MONADELPHIA. 19. SYNGENESIA.
17. DIADELPHIA. 20. GYNANDRIA.
18. POLYADELPHIA.

DICLINIA (a δὶς bis & κλίνη thalamus) duplex thalamus.
Mariti & Feminæ distinctis thalamis gaudent.
Flores masculi & feminei in eadem specie.
21. MONOECIA. 23. POLYGAMIA.
22. DIOECIA.

CLANDESTINÆ.
Nuptiæ clam instituuntur.
Flores oculis nostris nudis vix conspiciuntur.
24. CRYPTOGAMIA.

CLAS-

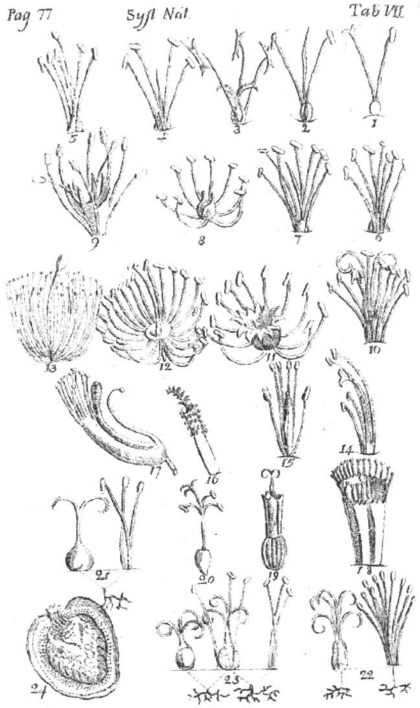

Figure 6.6. *(Linnaeus 1767, p. 22) (source: gallica.bnf.fr); (Linnaeus 1748, plate VII, p. 77) (source: Wikipedia)*

6.4.2. *Various approaches between description and classification*

Shortly after the publication of Linnaeus' main works, Bernard de Jussieu (1699–1777) designed a botanical garden for King Louis XV at Trianon, which was intended to reflect the classification of plants: small areas would represent a genus, genera would be grouped by family, etc. Similarity and difference criteria then arose, very concretely, in terms of the harmony of the garden. De Jussieu understood that within the same family, in addition to Linnaeus' criterion of reproductive organs, it was necessary to include a criterion corresponding to the morphological type of the flower: this would be the organizational plan. Without completely rejecting the Linnaean system, Jussieu adopted a natural method, taking all the characteristics and relationships between plants into account (Lamendin 2013, pp. 57–58). The classification would eventually consider both the Linnaean criteria and other criteria related to plant form. This was the work of Antoine-Laurent de Jussieu (1748–1836), who left behind a classification system that, to this day, remains the basis for current approaches and strives to integrate a set of morphological traits.

Michel Adanson (1727–1806) was born in Aix-en-Provence and lived in Paris, where his father, chapel master to the archbishop, Monsignor Charles-Gaspard de Vintimille du Luc (1655–1746), had followed his patron and teacher. He received a proper education owing to the archbishop's protection and was introduced to science by the masters of the King's Garden (Le Monnier, Rouelle and Bernard de Jussieu) and by the most famous naturalist of the period, Réaumur. After a trip to Senegal, Adanson was introduced to Louis XV in 1754 by Bernard de Jussieu and became Royal Botanist at the Trianon Gardens (Bailly 1992, p. 144). His life's work was *Familles des plantes* (Plant Families), in which he attempted to classify plants into families on the basis of 65 different criteria so as not to overlook any and achieve the most natural classification possible. Adanson thus established 58 families of plants on the basis of the results obtained from his 65 systems or criteria. The families described are therefore the result of a convergence of elements. However, the result, the coordination of convergent traits, is complex and difficult to use.

Antoine-Laurent de Jussieu therefore returned to the essential and permanent traits of the various groups of plants. He reintroduced a hierarchy of traits into his classification, a hierarchy that had been abandoned by Adanson. André Bailly wrote the following on this subject:

> It is by taking this hierarchy into account that families must be defined, when external signs of kinship are not evident. This is how he carried out his work on buttercups in 1773, which made it possible to define the family in a few sentences, whereas Adanson wrote six pages of description, in which some of the traits noted by Jussieu were not mentioned. (Bailly 1992, p. 174)

Finally, we should mention Jean-Jacques Rousseau, for whom botany was one of the major interests of his life. His *Lettres élémentaires sur la Botanique, à Madame Delessert* (Elementary Letters on Botany, to Madame Delessert) were published in 1802, with additional explanations by Antoine-Laurent de Jussieu and René-Just

Haüy. Rousseau thus emerged as someone who popularized botany, a science that he knew and practiced. Like others, he condemned Linnaeus' so-called artificial method and adopted the Jussieu brothers' project of natural description and classification. Rousseau, an avid follower of natural classification, proposed an almost pre-Romantic vision of botany that went beyond the science of the Enlightenment. This was even more true of Jacques-Henri Bernardin de Saint-Pierre (1737–1814), who was the intendant of the King's Garden in 1792 and dreamed of botanical geography. However, Rousseau and Bernardin had differences, particularly with respect to Bernardin's view of the limitations of studying plants outside their natural environment, whereas Rousseau emphasized the scientific role of herbariums (Drouin 2001, p. 510).

6.5. Conclusion

So much has been written on the subject discussed here that it would be useless to attempt to summarize it all in a few pages. We have therefore decided to focus on a few important but not well-known points in the history of botany from the 16th to the 18th centuries. We have chosen to begin by exploring the influence of Renaissance botanists, from Matthiole, who emphasized the medicinal uses of plants, to Cesalpino, who was the first to attempt a classification system. Following these considerations on botany in the 16th century, we discovered Peiresc, a correspondent of Charles de l'Écluse (he sent him seeds and plants) and a creator of gardens, where he acclimatized Mediterranean and Oriental plants. We believe that the role of gardeners in the period following the Renaissance and during the classical age should be reassessed, as should the role of observation and journeys of discovery, which became increasingly frequent during the classical age.

A second foundation for the development of botany and, above all, plant classification, appeared in the renewal of logic, first with Pierre de la Ramée, then a century later with the logical thinkers of Port-Royal. It is essential to understand here that *Port-Royal Logic* was one of the most widely read books of the 18th century and that scientists, even if they rarely cited it, were all familiar with it. Although it cannot be proven conclusively, it is quite evident that this book had a strong influence on the subsequent development of natural history.

The third foundation of this history of observation and classification of living beings, from the Renaissance to the 18th century, brings together the few steps that, from de Tournefort to Antoine-Laurent de Jussieu, established botany as a science, with its methods of description and classification. All the efforts of 18th-century botanists made it possible to incorporate Linnaeus' brilliant discoveries on the criteria of plant sexuality into a more natural method that also incorporated other criteria and aspects essential to plant species, which was completed with the work of Antoine-Laurent de Jussieu.

6.6. References

Aristotle (1883). *Histoire des animaux*, volume II, J. Barthélemy Saint-Hilaire (trans.). Hachette, Paris.

Aristotle (1948). *Métaphysique*, volume I, J. Tricot (trans.). Vrin, Paris.

Aristotle (1997). *Les topiques*, J. Tricot (trans.). Vrin, Paris.

Arnauld, A., Nicole, P. (1981). *La logique ou l'art de penser, dite logique de Port-Royal*. Vrin, Paris.

Bailly, A. (1992). *Défricheurs d'inconnu. Peiresc, Tournefort, Adanson, Saporta*. Edisud, Aix-en-Provence.

Bauhin, C. (1623). *Pinax Theatri Botanici*. König, Basel.

Bertrand, A. (2018). Présentation. La fleur mise à nu par les botanistes même. *Cahiers philosophiques*, 2, 153, 79–81.

de Caritat de Condorcet, M.-J.-A.-N. (1847). Éloge de M. le cardinal de Luynes. In *Œuvres de Condorcet*, tome III. Firmin/Didot, Paris.

Chansigaud, V. (2009). *Histoire de l'illustration naturaliste. Des gravures de la Renaissance aux films d'aujourd'hui*. Delachaux/Niestlé, Paris.

Drouin, J.-M. (2001). Rousseau, Bernardin de Saint-Pierre et l'histoire naturelle. *DixHuitième Siècle*, 33, 507–516.

Drouin, J.-M. (2008). Tournefort (1656-1708) : un botaniste cartésien ? In *Peuton classer le vivant ? Linné et la systématique aujourd'hui*, Prat, D., Raynal-Roques, A., Roguenant, A. (eds). Belin, Paris.

Duport, D. (2001). La variété botanique dans les récits de voyage au XVIe siècle une glorification du Créateur. *Revue d'histoire littéraire de la France*, 101(2), 195–212.

de l'Écluse, C. (1601). *Rariorum plantarum historia*. Christophe Plantin/Ioannem Moretum, Anvers.

Foucault, M. (1966). *Les mots et les choses*. Gallimard, Paris.

Gérard, L. (1871). *Précis de l'histoire de la botanique*. Guérin, Paris.

Hanomou, A. (2008). Les voyages du Père Charles Plumier (1646-1704) : des dessins de la faune et de la flore des îles françaises d'Amérique (Martinique, Guadeloupe et Saint-Domingue) à la fin du XVIIe siècle. *Bulletin d'histoire et d'épistémologie des sciences de la vie*, 15(1), 51–71.

Hoquet, T. (2007). *Buffon/Linné. Éternels rivaux de la biologie ?* Dunod, Paris.

Hoquet, T. (2008). Linné et les systèmes : un commentaire des *Fondements botaniques*, section II (1736). In *Peut-on classer le vivant ? Linné et la systématique aujourd'hui*, Prat, D., Raynal-Roques, A., Roguenant, A. (eds). Belin, Paris.

de La Ramée, P. (1555). *Dialectique*. Wechel, Paris.

Lamendin, H. (2013). *Les de Jussieu, une famille de botanistes aux XVIIIe et XIXe siècles*. L'Harmattan, Paris.

Le Bovier de Fontenelle, B. (1740). Éloge de M. de Tournefort. In *Éloges des académiciens*. Van der Kloot, La Haye.

Legré, L. (1900). L'indigénat en Provence du Styrax officinal ; Fabri de Peiresc et Pierre Pena. *Bulletin de la Société Botanique de France*, 47(4), 401–414.

Lherminier, P. (2021). Le genre, maître-mot du classement. *Bulletin d'histoire et d'épistémologie des sciences de la vie*, 28(1), 47–64.

Linnaeus, C. (1736). *Fundamenta botanica quae majorum operum prodromi instar theoriam scientiae botanices per breves aphorismos tradunt*. Schouten, Amsterdam.

Linnaeus, C. (1748). *Systema Naturae, II*, 6th edition. Salvius, Stockholm.

Linnaeus, C. (1767). *Systema Naturae, II*, 12th edition. Salvius, Stockholm.

Lister, M. (1873). *Voyage à Paris en 1698*. Société des Bibliophiles, Paris.

Mabillon, J (1691). *Traité des études monastiques*. Robustel, Paris.

Magnin-Gonze, J. (2015). *Histoire de la botanique*. Delachaux/Niestlé, Paris.

Matthiole, P.-A. (1572). *Commentaires sur les six livres de Dioscoride*, J. des Moulins (trans.). Rouille, Lyon.

Peiresc, N.C.F. (1890). Lettre 41 à Dupuy. In *Lettres de Peiresc*, volume 2. Philippe Tamizey de Larroque/Imprimerie nationale, Paris.

Peiresc, N.C.F. (1896). Lettre 132 à Valavez. In *Lettres de Peiresc*, volume 6. Philippe Tamizey de Larroque/Imprimerie nationale, Paris.

Peiresc, N.C.F. (1898). Lettre 411 à Clusius. *Lettres de Peiresc*, volume 7. Philippe Tamizey de Larroque/Imprimerie nationale, Paris.

Perru, O. (2012). *Hommes d'Église et Sciences au XVIIIe siècle : vers une harmonie entre Raison, Nature et Création ?* Vrin, Paris.

Planchon, J.-E. and Planchon, G. (1866). *Rondelet et ses disciples ou la botanique à Montpellier au XVIe siècle*. Typographie De Boehm, Montpellier.

Plumier, C. (1693). *Description des plantes de l'Amérique*. Imprimerie royale, Paris.

Raynal-Roques, A. and Roguenant, A. (2008). Nommer et classer les plantes : un parcours de trois siècles. In *Peuton classer le vivant ? Linné et la systématique aujourd'hui*, Prat, D., Raynal-Roques, A., Roguenant, A. (eds). Belin, Paris.

Rousseau, J.-J. (1802). Lettres élémentaires sur la botanique, à Madame Delessert. In *La botanique de Jean-Jacques Rousseau*, Rousseau, J.-J. (ed.). Louis, Paris.

von Sachs, J. (1892). *Histoire de la botanique du XVIe siècle à 1860*, H. de Varigny (trans.). Reinwald, Paris.

Selosse, P. (2004). Prinzipien der botanischen Nomenklatur in der Renaissance. *Linguistica Antwerpiensia*, 3, 133–152.

Selosse, P. (2008). Les concepts de genre et d'espèce à travers l'évolution du modèle logique de la définition, de l'Antiquité à Linné. In *Peut-on classer le vivant ? Linné et la systématique aujourd'hui*, Prat, D., Raynal-Roques, A., Roguenant, A. (eds). Belin, Paris.

Synowiecki, J. and Noûs, C. (2020). De la plante comme un homme. Penser la sexualité végétale au XVIIIe siècle. *Dix-huitième siècle*, 52, 453–470.

de Tournefort, J.P. (1694). *Elemens de botanique, ou Métode pour connoître les plantes*. Imprimerie royale, Paris.

de Tournefort, J.P. (1700). *Institutiones rei herbariae*. Imprimerie royale, Paris.

Vaillant, S. (1718). *Discours sur la structure des fleurs, leurs différences et l'usage de leurs parties*. Vander, Leiden.

7

Models, Systems, and Metaphors for Living Beings, from Descartes to Barthez

Paolo QUINTILI
Università degli Studi di Roma Tor Vergata, Rome, Italy

7.1. Metaphor for life in the 17th century: the mechanism

As Roger aptly noted in a classic book, at the end of the Great Century and at the beginning of the Age of Enlightenment, in the field of life sciences, "everything that matters is mechanistic" (Roger 1963, p. 207). The success of this approach to the phenomena of life throughout the 17th century was very difficult, but it was not solely due to Descartes, the "prince" of mechanistic philosophers. The reality is much more complex. Until the time of Leonardo da Vinci (1452–1519), physicians continued to be divided into two camps, which were not always opposed. On the one hand, there were the followers of Galen of Pergamon (129–201 CE), the Galenists, with various theoretical and practical variations; on the other hand, there were the followers of Aristotle (*De part. animalium*, *Hist. an*), who were in the majority. What united the community – but was also, at times, a source of division – was respect for Hippocrates' wisdom, who was held in high esteem by both sides. It was a scholastic, scholarly form of medicine, far removed from what we now call "clinical medicine". For a millennium or more, European and Arab physicians were guided and divided by this scholastic medicine of the four "humors" (blood, phlegm, yellow bile and black bile, or atrabile), which represented the great order of the physical macrocosm in the microcosm of living beings. Since the pre-Socratics, there have been four elements: air, fire, earth and water. In a medical context, these

History of Life Sciences and Medicine 2,
coordinated by Céline CHERICI.

elements were also divided into four more general "qualities" that became functional: heat (warmth) and humidity (wetness), cold (coldness) and dry (dryness) (Cosmacini 1997; Grmek 2014a). The disorder (*discrasìa*) of *Isonomia*, in other words, the balance, harmony (*eucrasìa*) of principles in the mixture of humors, is the cause of all disease, determined by a bad dyscrasia, by excess or deficiency (in the hot/wet or cold/dry qualities). Until the 17th century, almost all biologists and physicians – Aristotelians (scholastics), Galenists and Hippocratics – were, in different ways, *finalists*, following Aristotle (2014):

> For in the works of nature, it is not chance that is present, but that which is "in view of something", and it is mainly found there. [25] And the purpose for which a being has been formed or has come into being occupies the place of beauty. (*De part. animalium*, 645a, pp. 24–25)

The most general and widely held understanding of life, up until the Renaissance (inclusive), was therefore imbued with this profound finalist conviction, inherited and transformed by Galen in his theory of the "faculties" of the living body (primarily the human body[1]). For every being "endowed with a soul" (*empsychiché*), "to live" meant to maintain their own ends, to keep the harmony of their parts (solid and liquid) and with the universe of external beings. Every living being, every form of life, follows and obeys the immanent principle of its life (vegetative, sensitive, or rational "soul"). This form of teleologism took different forms but remained fairly consistent and unified in its fundamental principles until the 17th century. The metaphor that best represented and modeled this millennial finalist conception was the astrolabe, with its correlative representation of the zodiac man (see section 4.1). The macrocosm of the universe of "stars" (including the planets) is ordered like a great animal, whose life is governed by the *logos* of divine providence (*prònoia*), which ensures sympathy among the different parts of the cosmos (man and the Earth constitute the center of reference) and, through a kind of pre-established harmony, *pre*determines the balanced order (*Kosmos*), the right ends, positions and reciprocal relations. The microcosm of living beings, primarily man, is a small reflection of this great living animal that is the *Kosmos*, animated by the *logos* that permeates it. The metaphor – of Stoic and Platonic-Aristotelian origin – is perfectly reversible. The macrocosm was also a "macroanthropos"[2], a being that contained within itself, in an eminent way, the typically human qualities of intelligence, spirit and wisdom.

1 Galen (1994) in his treatises: *On the Usefulness of the Parts of the Human Body* (Vol. 1) and *On Natural Faculties* (Vol. 2).

2 Cosmacini (1997, p. 238), referring to Vesalius (1543) and his anti-Galenic revolution: "The universe was a macro-anthropos and man a microcosm".

Another characteristic of pre-Cartesian biology, up until the Renaissance and during the first half of the 17th century, whether it be Galenist or anti-Galenist, Aristotelian or anti-Aristotelian, is the presence of a *soul* – or its equivalent: spirit, quality, essence, etc., more or less immaterial – in this micro- and macrocosmic machine. Until the iatrophysics of Giovanni Alfonso Borelli (1608–1679), a disciple of Galileo in medicine, the physical machine of the body was conceived as being regulated by an "effective cause of movement and therefore of life [which] is the *soul*, which explains its action on the organism through the nerves that set the muscles in motion" (Borelli 1680–1681, vol. 1, p. 63). While introducing new concepts and practices in medicine – the "spagyrical" principles (mercury, salt and sulfur) as the "bases" of living nature, the use of chemical substances such as medicines and antimony – the great alchemist and "magician" Paracelsus (1493–1541), described as "*Medicus gloriosus*, as pretentious as he was provocative" (Cosmacini 1997, p. 239), believed in this *animating* essence, contained even in inorganic substances. Paracelsus was not innovative in proposing an alternative model of life to the traditional model, nor was he closely related to the new science of the Galilean scientific revolution, apart from his rejection of all authority (Aristotle, Galen, Avicenna and the schools[3]). The tradition of the "world soul" – of Stoic and Neoplatonic origin – was present and active right up until the height of modernity and the scientific revolution[4].

The first revolutionary shift from this ancient model of living beings, which went beyond the astrolabe metaphor and subsequently shattered the macrocosm-microcosm equivalence, came with the mechanistic philosophies of René Descartes (1596–1650) and Galileo Galilei (1564–1642). In his brief reflections on medicine, which was set out in his *Dialogue Concerning the Two Chief World Systems* (Galilei 1632), Galileo was the initiator – or rather the inventor – of a general approach to nature that readily dispensed with the presence of spirits within it. With his principle of the autonomy of the language of nature from theology, which Popper (2002) aptly defined as the "demarcation criterion" between science and non-science (not necessarily "pseudoscience", as can be seen in the case of Paracelsus), Galileo ushered in an era of true revolution with his mathematical approach to natural phenomena, which has been adopted by science up to the present day. His view of nature as an object was based solely on *measurable facts*, "proven and reproven", dissociated from any speculative, theological or metaphysical concerns, with the intention of establishing universal physical and mathematical laws capable of developing controllable and reproducible experimental *protocols*. This method,

3 Pagel (1984) takes a more nuanced view of the scientific revolution.

4 Gregory (1955, 2000) on overcoming this metaphysical view of the world.

rightly called "Galilean", is opposed both to the rigorous formalism of the Aristotelians and to the speculations (mixed with operational pragmatism) of Renaissance physicians, as well as the first iatrochemical school inspired by Paracelsus (van Helmont (1580–1644) 1648; Grmek 2014a). Galileo was deeply convinced of the profound unity of the natural world, from the inorganic to the organic, and his *Platonism*, which had been assimilated in Renaissance circles in Florence, had given him the belief that *numbers* and mathematics were the language in which "the book of nature is written" (Galileo 1632, 1692). This implied that, from the inorganic to the organic world, the connecting fabric that links all phenomena is *quantity*, as well as the laws of motion and inertia, which are the same for all beings and everywhere in the universe. On the second day of his *Dialogue* (Galileo 1632, 1692), Galileo explains the movement of animal joints by the simple laws of mechanics and by the unitary principle that underlies them:

> If, in the movement of a solid body, one of its extremities remains in the same place, its movement can only be circular; now, when an animal moves a limb, it does not separate it from the limb adjacent to it, so this movement is necessarily circular. (ibid., p. 267)

It is not that Galileo denied the existence of a "soul" in all living beings (ibid., p. 266), but their structure and "nature" (their behavior) always obey the same laws of general physical nature (Castiglioni 1942; Barcaro 1982; Grmek 2014b; Canguilhem 2002, pp. 37–50). This mechanistic approach is positively embraced by Descartes and carried to extremes. However, Descartes argues directly against what he considers to be Galileo's "empirical" method, which he believes does not lead to *truly* universal laws of nature, whether they be living or non-living (Koyré 1937; Roger 1963; Aucante 2006, pp. 48–49, 94–97[5]). The new mechanistic approach to the world of life needs to be based on a new method, on general models of a systematic and deductive nature and, at the same time, on a *metaphor* of the living machine (and the body) that summarizes all phenomena within the same framework. This would ultimately be the model of the hydraulic machine, "the fountain" (Descartes [1662/1664] 1996). In his *Treatise on Man and the Formation of the Fetus* (Descartes [1662/1664] 1996), Descartes uses this metaphor while being

5 Descartes (2000); see the letter to Mersenne dated October 11, 1638, in Descartes (1949, p. 802): "Without considering the first causes of nature, he [Galileo] only sought the reasons for a few particular effects, and thus he built without foundation", because, according to Descartes, "true science" can only be deductive, that is to say, the result of clear and distinct demonstrations and deductive reasoning (following the example of geometry).

aware of the need for a "model" to describe the true functioning of the machine of the body:

> We see clocks, artificial fountains, mills, and other similar machines, which, though made only by men, nevertheless have the power to move themselves in many different ways; and it seems that I cannot imagine many powerful movements in this one, which I assume is made by the hands of God, nor attribute so much artificiality to him that you would have reason to think that there could be even more. (Descartes [1662/1664] 1996, p. 120[6])

Descartes also mentions clocks and water mills in this passage, but in his later writings, he most often uses the image of a fountain or a complex hydraulic machine, even to describe the movements of the animal spirits that animate the muscles[7]. In his other posthumous works on anatomy (*De la formation du fœtus, Excerpta anatomica, Primæ cogitationes circa generationem animalium*), Descartes expressed his position as both epigenetic and mechanistic – animals are formed by the mechanical juxtaposition of parts according to the laws of motion and heat – accompanied by a *cardiocentrism* that was the result of his adherence to Harvey's (1628) thesis on blood circulation, in which the heart is the "sun" of the bodily system of every living being (Descartes 2000, pp. 123–157). Latronico (1955, pp. 41–43) emphasized the *collective* nature of the "discovery" of circulation, which is actually the result of the work of several contemporary anatomists and researchers (Colombo, Servetus, Sarpi, Fabrici d'Acquapendente and later, Harvey). Both Harvey and Descartes thus definitively broke away from the humoral medicine paradigm, opting instead for a "solidist" physiology – living beings are a composition of solid organs coordinated and "in harmony" with one another by purely mechanical causes – based on the postulate that living beings are the product of the *fermentation* of seed particles – of different shapes and forms – which resemble each other to form each "part" of the machine, the heart first and the brain

6 On the "failure" of Descartes' entirely mechanistic biology, see Roger (1963, pp. 140–154); on the importance of the notion of "model" in Descartes and his productivity for "the framework of hypothetical-deductive sciences as they gradually developed from the end of the 17th century onwards", see Aucante (2006, pp. 428–429).

7 Ibid., pp. 130–131: "And truly, one can very well compare the nerves of the machine I am describing to the pipes of these fountain machines; its muscles and tendons to the various other devices and springs that serve to move them; its animal minds to the water that moves them, whose source is the heart, and the concavities of the brain are the openings". Clerselier also highlights Descartes' interest in hydraulic "automata" in his "Table" in the *Appendix* to his posthumous edition (Descartes 1664, pp. 212–215).

second[8]. Roger (1963, p. 144) asserts that "the ideal outcome clearly expressed by Descartes is mathematical embryology". The dynamics governing these phenomena are always modeled by the metaphor of the *hydraulic machine*, with the heart as the "motor" and source of heat[9].

In this cycle of the formation of living beings, the mechanical form of reasoning adopted by Descartes led him to imagine the existence of a *microscopic geometry* of particles, which, having different shapes and figures, could come together in different ways for each organ and each anatomical part (veins, arteries, vessels, tissues, etc.). For Descartes, this mechanism of living microscopic particles paved the way to a basic form of corpuscular *atomism* – which was similar to the much-criticized Democritus and his followers, Epicurus, Lucretius, the materialists – and did not fit well with the rest of Cartesian biology. In particular, the Cartesian physiological approach failed in the face of the problem of *sex* formation in the embryo. A supporter of the (Aristotelian) theory of the double seed in the *Primae Cogitationes*, and later in his posthumous writings and *Excerpta*, Descartes hesitated between several solutions as the cause of sexual difference – the position of the fetus in relation to the mother, the nature and purging of excrement in the womb and the pressure of the "humors". These "solutions" clearly showed the purely hypothetical nature of his research on generation. Descartes was therefore far from possessing a science of living beings that could claim the certainty and geometric evidence required by deductive physics. This was a shortcoming, which Descartes himself was very well aware of.

This point of inconsistency between experimental research (pragmatically *conjectural*) and general principles of deductive reasoning (clarity, certainty and evidence, which are impossible to find in the realm of living things), especially with

---

8 Ibid., pp. 256–257: "As soon as the heart begins to form in this way, the rarefied blood that flows out of it takes a straight course toward the place where it is freest to go, and that is the place where it forms after the brain; just as the path it takes begins to form the upper part of the main artery".

9 Regarding the difference between Descartes' cardiocentric system and that of Harvey, the former conceives of the heart not as a simple "hydraulic pump", but as a kind of water "boiler", producing propulsive force in the form of heat (Canguilhem 1971, pp. 112–116, 1977a, p. 172). However, he expresses reservations about the Cartesian "break" with Aristotelianism, which he does not consider entirely revolutionary in terms of finalism. For Descartes, finalism becomes "technological", because it is inspired by the technical purpose/functioning of a machine (created by God).

respect to the *conception of living matter*, is what critics used to attack Cartesian biology with great success[10].

7.2. Criticisms of the Cartesian mechanism: pre-existence of germs, iatrochemistry and animism

Another weakness of Cartesian biology, beyond embryology, was the question of substance *dualism* – *res extensa* (body = matter) and *res cogitans* (soul = thought) as ontologically distinct – and the repercussions that this metaphysical problem had in the life sciences, particularly in anthropology. In his *Treatise on Man*, Descartes left unresolved – or rather did not address at all – the argument concerning the combination of the two substances, which he had announced at the beginning of his treatise. This is the problem of the "third substance", the vast field of the *union* of the soul and the body, as well as its phenomena:

> And I must describe to you, first, the body separately, then the soul separately; *and finally, I must show you how these two natures must be joined and united* to compose men who resemble us. (Descartes 1996 [1662/1664], pp. 119–120, italicized by Descartes himself)

However, the description of the "separate soul" and the modalities of "joining" these two "natures" are missing from the posthumous treatise. The numerous aporias left open by this proposition sparked heated debates that would continue for more than two centuries – and even to this day – in European scientific and philosophical circles, in what came to be known as the mind-body problem (MBP) (Carter 1983; Crane and Patterson 2000; Almog 2002; Aucante 2006; Lagerlund 2007; Quintili 2009a, 2024; Westphal 2016).

It was not simply a question of "rational psychology" or "empirical psychology" – to use scholastic terminology – concerning man and his "soul". This difficulty also extended to the representation and conception of the soul of *animals*. The question is whether animals are merely "pure machines" – as Descartes argued, since, strictly speaking, the soul is nothing but thought (*cogito*) and language, which animals lack

10 Roger (1963, p. 153): "What is at stake, it seems to us, rather than a question of scientific method, is the way in which living matter is perceived. Descartes wanted to apply the imagination of a physicist and geometer to life, or rather the imagination of a mechanic. He played with particles as if they were gears. No doubt he thought that man would one day intervene in these intricate mechanisms, or even build a machine as admirable, but ultimately as simple". Descartes' descriptions are not *just* "images", but rather metaphors/models necessary for the "system", without which Descartes could not have built his physics.

– or whether animals are also endowed with cogitative and rational faculties and therefore have a soul (spontaneity), as argued, for example, by the Italian physician Nicolas Rorarius, Nuncio of Pope Clement VII at the court of Hungary, who lived in the first half of the 16th century and whose doctrine (Rorarius 1566) is recounted (against Descartes) by Pierre Bayle in his *Dictionnaire historique et critique* (Historical and Critical Dictionary):

> He has written a work that deserves to be read. In it, he sets out to not only show that animals are rational beings, but that they also use reason better than humans. (Bayle 1697, Vol. 4, p. 955)

This marked the beginning of the *Querelle sur l'âme des bêtes*[11] (Cohen-Rosenfield 1940; Markovitz 1987; Marcialis 1993, Torero-Ibad 2009). Without going into the details of this complex issue (which are mainly metaphysical), we can simply say here that critics mostly attacked the empirical and experimental evidence for the falseness of the Cartesian thesis of the animal machine. Animals are endowed with a "soul", which can then be conceived in several possible ways, even "materialistically". This is the case with the famous physician Willis (1672), as well as Gassendi (1658) and Leibniz (1710), cited in Leibniz (1890), to mention only the most significant ones. This was followed by a great debate that gave rise to the new "organicist" conception of animals during the Enlightenment, notably from de La Mettrie ([1750] 1987) and Condillac (1755), which will be discussed further below.

Once it had been established that mechanism was the most rigorous way of explaining the nature of living beings but that the Cartesian model was insufficient and ineffective for this purpose, after 1670, with the dissemination of Descartes' posthumous work, several scholars, philosophers and physicians moved toward a *critical*, refined and improved *retrieval* of certain older doctrines: iatrochemical, animistic and even Aristotelian doctrines from the previous century, which had been discredited by the best natural philosophy of the scientific revolution (Galileo, Borelli, etc.). These doctrines were updated and used against Cartesianism or merged with a *new mechanistic way of thinking*. What was missing from the mechanism of living beings, which was very poorly conceived by Descartes, was what would *animate* it, namely, giving it form – the problem of form and *formation* once again became the focus of the argument – and keeping it alive, without necessarily always resorting to the action of Providence. Obviously, when faced with the failure (and challenge) of Cartesian biology, the scientific community's first impulse was to integrate classical metaphysics – and even Augustinian theology – in a renewed way into the science of a priori mechanism.

11 Quarrel over the souls of animals.

After 1650–1670, the dates of Descartes' death and the spread of his philosophy, what Hazard (1938) aptly called "the European crisis of conscience"[12] began to take root. As ideas flowed, old systems faltered, metaphysical certainties were questioned once again and it seemed that a return to the "old world" might be the wisest solution. The first waves of certainty came from experimental research, with new discoveries that had not affected Descartes' doctrine: the discovery of eggs in viviparous animals by Regnerus de Graaf (1641–1673) (de Graaf 1668) and others (Johann van Horne, 1621–1670) and the discovery of spermatozoa by Antoni van Leeuwenhoek (1632–1723) (van Leeuwenhoek 1687). The discovery of "atomic" elements that make up the "seed" of male and female animals meant that we had touched upon the basic building block of life. Following this new perspective, which was made possible in part by the use of a microscope (van Leeuwenhoek), attempts were made to *observe* the production of living beings directly. The first efforts in this direction produced a new biological "system", that of the *pre-existence of germs*.

Between 1668 and 1674, the Dutch naturalist and microscopist Jan Swammerdam (1637–1680), the Parisian physician Claude Perrault (1608–1680), the Italian physician and anatomist Marcello Malpighi (1628–1694), and the Oratorian philosopher and theologian Nicolas Malebranche (1638–1715) almost simultaneously developed the doctrine that the germs of every living being already existed and had been created and formed by God himself – in the egg ("ovism") or in the "spermatic animalcule" ("animalculism") – and had been interlocked since the beginning of time, since the creation of the world, for the human species: in Eve's ovaries or in Adam's testicles. Perrault (1680), Malpighi (1673) and Malebranche (1674) "see" this pre-existing germ in the egg; Swammerdam (1669), van Leeuwenhoek (1687) and others after him in the animalcules. The Cartesian mechanism, therefore, is merely the instrument, the technique of development or "evolution" over time, of a being that already exists, created in germs, since the beginning of time, by God. The pre-existence of germs should not be confused with classical preformation, a medical-metaphysical hypothesis defended by Liceti (1577–1657), Parisano (1567–1643) and other naturalists of the Baroque era, whose principle is that the form of living beings, their true formation, is given prior

12 "What a contrast! What a sudden change! Hierarchy, discipline, order enforced by authority, dogmas that strictly regulate life: this is what men of the 17th century loved. Constraints, authority, dogmas: this is what men of the 18th century, their immediate successors, detest. The former are Christians, and the latter are anti-Christians [...] The majority of French people thought like Bossuet; suddenly, the French think like Voltaire: it is a revolution" (Hazard 1938; p. 19).

to birth in the body of the progenitor and by its "soul" and which had been successfully contested by Harvey and Descartes in the early 17th century (Roger 1963, pp. 325–333; Grmek 2014b). The system of pre-existence had two advantages: it was perfectly in line with Christian orthodoxy, which was threatened by the spread of Descartes' philosophy in certain "versions" of materialist or monist inspiration (Spinoza and Spinozism) (Vernière 1954; Moreau 2003; Vandewalle 2011; Andrault 2018; Lodoli and Quintili 2020), *infra*, section 6.3); it proposed a biological model (the precreation of living forms by God) as an alternative to mechanistic epigenesis, a dangerous mechanism because it implied an autonomous power of generation *within* matter, owing to the simple "forces" of heat and movement; and in certain variations (chemism), through the influence of a "soul" that acts and directs (Duscheneau 1974). Mechanism does not "form" living beings; it merely develops them[13]. Proponents of this doctrine claimed to have "seen" and "experienced" the existence of this germ, formed in very small quantities in the seed of living beings: Swammerdam in the nymphs and caterpillars of insects, Malebranche in the grain of a tulip, Malpighi and Perrault in a fertilized chicken egg and van Leeuwenhoek in the spermatozoid itself. The latter even provided a very suggestive and imaginative illustration (section 4.5). This is one of the most striking testimonies to the errors that can be made by a strong "scientific ideology" when using technical means that are still imperfect, even within the framework of a "rational" scientific theory, and despite the evidence of the facts[14]. However, the pre-existing germ "seen" in the spermatozoon is also one of the major blunders in the history of biology, linked to the use of the microscope (section 4.5; on microscopy and its errors, see Solinas (1967) and Croft (2006)).

Countless new aporias of the pre-existence of germs thus arose against the systems of living beings at the beginning of the 18th century:

– the extreme, even infinite, smallness of the first germs;

– the nesting of these germs within the prototypes of each animal species, which must therefore have been gigantic in size at the outset;

13 "What remained of Cartesian science, apart from mechanism, a mutilated mechanism that was left with nothing more than the task of making living machines work, which it was incapable of 'constructing'? And this incomplete mechanism was already in danger of becoming an act of faith" (Roger 1963, p. 443). On Malebranche, see, in particular, Andrault (2016).

14 In this historical case, the concept of "scientific ideology" can be used to refer to a set of pre-existing conceptual representations that are used to legitimize a view of the natural world, governed by dominant and shared religious (and, later, political) beliefs (Canguilhem 1977b).

– the "dynamism" inherent in living beings, exemplified by the regeneration of limbs in certain simpler animals (lizards, freshwater polyps (Duchesneau 1998; Andrault 2014));

– the phenomena of heredity, hybrids, monsters, etc.

Where is the "hand of God" (and what is it doing) in these phenomena? Above all, what is the purpose of a *pure mechanism*, limited to a few operations (laws of impact, inertia, extension, impenetrability and spring), in explaining the formation of an insect wing or a blade of grass? The revival of the tradition of Jan Baptista van Helmont (section 4.1) – the "spiritualist movement" (Roger 1963, p. 97) in medicine during the Baroque period – and of the iatrochemical school, inspired by Paracelsus, associated with the spread of the *Spinozist model* (Duchesneau 1974; Quintili 2013; Andrault 2014), represents the first systematic response to these new difficulties of mechanism in the life sciences, generated by its alliance with the ideology of the pre-existence of germs.

The first person to inaugurate this rediscovery of the vital forces of animism was the German chemist and physician Georg-Ernst Stahl (1660–1734), who founded a school of his own in Halle and Berlin – at the court of Friedrich-Wilhelm I, where he practiced his profession – and which greatly interested encyclopedists, notably Baron Paul-Henri Thiry d'Holbach (1723–1789), a materialist philosopher of German origin. D'Holbach translated several of Stahl's works on mineralogy into French and for the *Encyclopédie* (Stahl 1766, 1771). This unusual alliance between animism and materialism (or "Spinozism") should come as no surprise. Stahl's central thesis, inherited from van Helmont, acknowledged that to regulate and direct the formation and subsequent operations of an organic body, that is, to explain what life itself is, one must grasp the *Arche*, the formative principle (or "Artisan") of this life – the *Archeus Faber*, according to van Helmont's expression, "a principle of its movements, an exciter and an internal director of its generation" (van Helmont 1648, Chapter VII, p. 40), cited in Roger (1963, pp. 99–100)). We cannot understand the purpose, the "end for which" (Aristotle) the organism is constituted, without considering its physical action through the (chemical) substances it uses and releases.

According to Stahl, this *arche* is the "soul" that accomplishes its work in and on the body (Stahl 1706, cited in Stahl 1860, 1708). However, this animism that "vitalizes" the organic machine – which cannot function by the simple laws of inertia, shock, etc. – has implicitly materialistic connotations, which Stahl always denied but did not escape a very attentive critic such as Leibniz (1768, Vol. 2) in his *Animadversiones circa Assertione aliquas Theoriæ medicaæ veræ Clar. Stahlii* (Leibniz 1709). The introduction of a "vital principle" that was beyond the order of the mechanical laws of nature – which Stahl likened to chance – made the science of living beings incomprehensible, according to Leibniz, and, above all, turned the soul into *some sort of body*, confusing it with the body, in short, leading to materialism[15]. It is from this perspective that Stahl's medicine and chemistry – precursors to what we now call *psychosomatic medicine* (Hoffman 1996); on the correspondence and controversy between Leibniz and Stahl, see (Leibniz 2016)) – were received a few decades later by Baron d'Holbach and the encyclopedists.

Other authors shared Stahl's criticism (and dissolution) of mechanism, inspired by or belonging to schools that were more or less anti-Cartesian or neo-Platonic, who reintroduced "souls" into matter to define living beings: the Cambridge philosophers More (1614–1687) and Cudworth (1617–1688), the journalist and theologian (an enemy of Bayle) Le Clerc (1657–1736) and, above all, the Dutch physicist Nicolaas Hartsoeker (1656–1725), who lived and worked in France for a long period of time. In his *Conjectures physiques* (Hartsoeker 1706) through *Éclaircissements* (Hartsoeker 1710), *Suite des Éclaircissements sur les Conjectures physiques* (Hartsoeker 1712) and then in his *Cours de physique* (Hartsoeker 1730), Hartsoeker demonstrated the new scientific spirit that ushered in the Age of Enlightenment. It was a spirit of *conjecture* (a new term that would be widely adopted in the future), which was no longer a mere synonym to the empiricist and fideist *skepticism* that had, for example, characterized Gassendi or Bayle (God always knows more than human reason can understand), nor a mere opponent of mechanistic dogmatism, but rather introduced a definitive affirmation of the experimental method in the life sciences, namely, a *phenomenalism* that closely followed the in vivo experiences of nature while at the same time, not severing its ties with metaphysical speculation, that of the *Mirabilia* that fill the pages of the *Conjectures*.

Initially, a believer in the pre-existence of germs, Hartsoeker had the courage to change his mind and became a vitalist when new experiments and discoveries concerning the regeneration of crayfish legs occurred. According to him, this type of phenomenon demonstrated the presence of an *intelligence* or *force* that guided the process of organ formation and *reformation*[16]. The same applies to medicine: it is

15 "The soul does not give the body any movement, any degree or direction of movement that is not the mechanical consequence of the previous states or movements of matter. To assert the contrary is either to change the soul into a body or to resort to inexplicable principles" (Leibniz 1768, Vol. II, *Responsiones ad Stahlianas observationes*, ibid., p. 150).

16 "Given that, according to Mr. Cheyne himself, the mechanical formation of the Animal is inconceivable and impossible, and since it is absurd to think that God creates something new in the production of each Animal; the mechanical formation of a single claw or paw is no less inconceivable, no less impossible, and no less absurd than to think that God creates something new. Thus, it is necessary to resort to some kind of Intelligence, whatever it may be, which resides in this Animal and repairs this loss as soon as it happens" (Hartsoeker 1717, p. 336).

this intelligence or force, acting in the organs and throughout the body, that "repairs" it – as Stahl also taught (King 1964, 1978; Oldroyd 1973; Debus 1977, 2014) – when it falls ill. Hartsoeker was accused of being a "Spinozist" because he considered the soul and the body to be simply "different qualities"[17], whereas Cartesians regarded the two as being ontologically distinct substances.

This paved the way for the affirmation of this organicist "neo-Spinozism" in biology (Vernière 1954), which would shape the philosophical perspective of the vitalism of the Enlightenment and the Montpellier School.

7.3. Spinoza and the "Spinozism" of the Enlightenment: the new life sciences

Upon closer inspection, the *philosophical basis* of Stahl and Hartsoeker's medical and biological doctrines was, in fact, inspired by Spinoza, even though Hartsoeker remained fairly faithful to classical deism, on the basis of the idea of a God who is the architect of the world. Stahl did not fail to include all the operations of the *arche* or the Principles of Life – the human soul – in his conception of God (the Supreme Distiller, *sive Natura*) without worrying about the philosophical "dangers" pointed out by the genius of Leibniz. The life sciences of the Enlightenment arose against this level of ambiguity and, consequently, against this new doctrinal *freedom* from the old dogmatic ties that had now been broken, whether they were Aristotelianism, Galenism or even successful Cartesianism – a victory that was short-lived – at the beginning of the 18th century.

The "minor physics" or "abstract physics" that Spinoza sets out in the second part ("De Mente") of his *Ethics* (Spinoza 1677, posthumous) is a summary of physiology, inspired by neo-atomism and materialism. Living bodies are aggregates of "individuals" (*corporis compositi*, "compound bodies") that are constantly changing and renew themselves internally through a dynamic exchange of interaction between the bodies themselves, between the inside and the outside, and between the organism and its environment. The decreases and then cessation of this renewal and exchange constitute the death of the entire organism[18]. Spinoza's conception of the living being is still, in many ways, based on the Cartesian

17 "Thought and extension may have nothing in common with one another, and yet belong to the same substance" (Hartsoeker 1710, p. 105).

18 "The object of the idea that constitutes the human soul is the body, in other words, a very specific mode of extension that exists in actino, and nothing more" (Spinoza 2020, pp. 179–197, propositions 13–19).

iatromechanistic view – there is a kinematic relationship between organic parts that follows the principle of inertia, the laws of impact, etc. – but, at the same time, it moves away from it in the field of this *physiological theory of the individual* – the body composed of small bodies – leading to a vitalist cosmology of "true life" (God/Life/Nature) that is similarly inherited from Jewish theological and scientific culture (Maimonides) and which is closer to Leibniz's contemporary doctrines on monads (Zac 1963; Duchesneau 1974; Bouveresse 1992; Macherey 1997; Vandewalle 2011, pp. 50–58; Andrault 2014). Spinoza's philosophy is undoubtedly not a form of vitalism, as he does not attribute the life of beings to an "arche" or principle, regardless of whether they are internal or external, this life being fundamentally inherent to them. However, this philosophy, centered on the primacy of the human body over the soul (the body being the object of the soul's ideas), inspired and promoted the diffusion of a form of *panpsychism* between the end of the 17th century and the beginning of the 18th century, a panpsychism that is also characteristic of Leibniz's metaphysical and biological thought[19]. Monads are species of metaphysical/living individuals on the basis of all physics, and all natural individuals, simple and/or composite, that populate the world, according to Spinoza, are *animated* to varying degrees[20]. This scholia on proposition 13 of Part II of *Ethics* paves the way and supports the development of animist and vitalist – anti-Cartesian – biological theories, which, in a sense, usher in a new perspective for Enlightenment philosophy in the field of life sciences.

The "Spinozist" system of life[21] was therefore the more or less direct and explicit source of inspiration for the first heterodox doctrines of the early 18th century. The Dutch physician Hermann Boerhaave (1668–1738), who was the mentor of the first materialist philosopher-physician of the Enlightenment, Julien Offroy de La Mettrie (1709–1751), was accused of Spinozism during a boat trip from Leiden to Amsterdam for daring to ask ministers of religion, who were in a rage against Spinoza, if they had ever read the books of this renegade Jew (Knoeff 2002, p. 21; Diderot 2004, p. 342). Boerhaave's view of life, which reconciles mechanism and chemistry while remaining iatromechanistic, is nevertheless marked by the same concern for complexity that characterized Spinoza's view. This is the main contribution that Spinoza and (above all) 18th-century Spinozism made to the

19 Bouveresse believes that Leibniz's panpsychism may have been inspired by Spinoza himself (Bouveresse 1992, pp. 7–8).

20 "For what we have shown so far is entirely common and does not apply more to humans than to other individuals, who are nevertheless all animated, although to varying degrees" (Spinoza 2020, p. 179, E2, 13, scholia).

21 The word "system" is used rather ambiguously in the case of Spinoza (Ibrahim 1990).

doctrines of life: the concern for complexity that was lacking in Cartesian science. An example of this is provided by Boerhaave's theory of generation, which occurs through the encounter, in the womb, between the egg and the spermatozoon "which enters through the dilated pores of the small membrane of the egg, which has become glandular" (Boerhaave 1739 and 1740 , I, p. 453), cited in Roger (1963, p. 315). This was a correct view, owing both to the microscopic discoveries of the late 17th century and to a methodological context that avoided Descartes' simplistic mechanistic approach.

His student, La Mettrie, adopted this "Hippocrates of Leiden" method. He first translated the *Institutions de médecine* (de La Mettrie 1739) and then embarked on research that would culminate in the Spinozist philosophy of *Man a Machine* (de La Mettrie 1748), where the author ends his work with the explicit statement: "Let us therefore boldly conclude that Man is a Machine; and that there is only one substance in the entire universe, modified in various ways" (de La Mettrie 1748, p. 108). Contrary to what the title of his masterpiece might suggest, living beings are not *pure* machines in the Cartesian sense but *animated components* of several living micromachines that coordinate with one another, each with its own sensitivity, and which together form the entire organic body (of the animal or the plant[22]). In doing so, de La Mettrie introduces a crucial concept into his system of living beings, that of "innate force" or, in other words, the *sensitivity of matter* present in each "individual" component of the organism:

> If I am now asked where this innate force in our bodies is located, I would answer that it clearly resides in what the elders called *parenchyma*, that is to say, in the substance of the parts themselves, apart from the veins, arteries, nerves – in a word, apart from the organization of the whole body – and that, consequently, each part contains springs within itself that are more or less active, according to the needs of that part. (Ibid., pp. 75–76)

This turning point in the history of life sciences, which was accompanied by the introduction of the neologism "organism" to refer specifically to a (living) compound of organs, was recorded at the time of de La Mettrie, in Diderot's

22 "Every tiny fiber or part of organized bodies moves by a principle that is unique to it, and whose action does not depend on the nerves [...]. Now, if this force can be observed even in pieces of fibers, the heart, which is a compound of singularly intertwined fibers, must have the same property" (ibid., pp. 74–75).

Encyclopédie and in other dictionaries[23]. This concept is disguised in the "Spinozist" article in *Encyclopédie*, which sheds decisive light on the issue. Diderot wrote the following entry:

> SPINOZIST, n. (*Gram.*) follower of Spinoza's philosophy. Ancient Spinozists should not be confused with modern Spinozists. The general principle of the latter is that matter is sensitive, which they demonstrate by the development of the egg, an inert body which, by sole means of graduated heat, passes into a sentient and living state, and through the growth of all animals, which in its beginning is but a speck, and through the nutritive assimilation of plants, in a word, of all substances that serve for nutrition, become a large sentient and living body in a large space. From this, they conclude that there is only matter, and that it is sufficient to explain everything; for the rest, they follow ancient Spinozism in all its implications. (Diderot 1751, Enc., XV, p. 474a)

The "Spinozist" thesis is, in reality, nothing more than a new biological and vitalist materialism that, beyond all previous "systems", is based on a rational *hypothesis*, namely, the conjecture of sensitive matter, of "innate force", on the basis of a fragmented structure of matter, made up of living atoms that aggregate and disaggregate to form organic architectures[24]. There are no longer any Cartesian-style *models* (the fountain), but *structures* subject to the laws of epigenesis, arising solely from this autonomous "force" that runs through them and *organizes* them. The path to a new foundation for the life sciences is indicated here for the generation of philosophers who would follow Spinoza's lesson in this new context, where the notion of life as an innate material force penetrates and acts – like Stahl's soul and Newton's attraction – in the de-finalized universe of living bodies. The door was

23 *Encyclopédie*, XI, p. 629a, contains the entries "Organic" ("Pertaining to an organ. The body is divided into organic and inorganic parts. See *Body* and *Organ*") and "Organization": "Arrangement of the parts that constitute living bodies. The first principle of organization is found in seeds. The *organization* of a body, once established, is the origin of the *organization* of all other bodies. The *organization* of solid parts is carried out by mechanical movements. See TLF online, under the entry 'Organism'": http://stella.atilf.fr/Dendien/scripts/tlfiv5/visusel.exe?12;s=4261513590;r=1;nat=;sol=1.

24 "Spinoza cannot, under any circumstances, provide any basis, "from within", for modeling; his legacy and tradition, forbidden from the outset by the nature of their object, therefore develop in an unprecedented way" (Ibrahim 1990, p. 123) (see also Thomson (1981, 2008) and Comte-Sponville (1990)).

open to this materialist vitalism, which would end up asserting itself successfully in the second half of the 18th century.

7.4. Diderot, the *Encyclopédie* and the vitalism of the Montpellier School

We must now follow the thread laid out by Diderot's work on the difficult path that also passes through the hidden underground streams of secret literature between the 17th and 18th centuries (Bloch 1982; Ibrahim 1990; Israel 2001). An anonymous work entitled *Telliamed, ou Entretiens d'un philosophe indien avec un missionnaire français sur la diminution de la mer, la formation de la terre, l'origine de l'homme, etc.* (Telliamed, or conversations of an Indian philosopher with a French missionary on the decline of the sea, the formation of the earth, the origin of man, etc.) secretly circulated in manuscript form from the beginning of the 18th century and was finally published in 1748 and 1755[2], after the author's death. It was written by the diplomat and historian Benoît de Maillet (1656–1738), who revolutionized the traditional biblical account of the history of the Earth and animals[25]. There was no "flood"; the Earth was naturally submerged in water from its inception, and living beings were not "created" but developed from the oceans – from water that gradually receded – and then spread over the land. Living beings are therefore the result of the same process of generation and gradual *metamorphosis* over time (an immense period of millions of years) as the formation/transformation of the globe. The title *Telliamed* is an anagram of the author's name (de Maillet), whose book was an immediate success. The image of the living world derived from *Telliamed* had a profound impact on the intellectual consciousness of scientific and philosophical circles in the first half of the 18th century (Bloch 1982, pp. 120–123, 236).

Buffon (1707–1788) and Maupertuis (1698–1759), the two leading naturalists of the time, developed their systems after this temporal perspective on the formation of life was broadened, later called "transformism", which had been inaugurated by *Telliamed* (read and appreciated later by Lamarck and Darwin). As a contemporary and elder of Diderot, d'Alembert and d'Holbach, Buffon published his *Histoire naturelle, générale et particulière* (General and Specific Natural History) in 1749, two years before the first volume of *Encyclopédie* (Diderot 1751), which marked a sort of baptism of the biological perspective on the new time scale of the expansion of living forms. The globe is a great "forge" that produces and destroys all forms

---

25 "It is still true that these foreign bodies, terrestrial or marine, are found even on the summits of our highest mountains. To attribute this marvel to the Flood is, according to our Philosopher, an untenable opinion" (Maillet 1748, I, p. xxvj; Benítez 1980).

and individuals during the seven "epochs of nature". Nature itself – *Mother Nature* – is a productive force made up of atomic elements shared by all living things, which Buffon called "organic molecules", immortal, indestructible atoms and infinite in number, which unite (life) or disintegrate (death), giving form to the living universe. These are the ultimate constituents from which all animal forms emerge (Buffon 1749, Vol. I and II). For the generation and morphology of living beings and to ensure the development of organic form in three-dimensional space, Buffon, a proponent of epigenesis, established the functional concept of the "internal organic mold", a material blueprint of the animal (or plant or other organism) that exists in nature, in which molecules encounter and follow in their aggregations, completing the epigenetic process (Roger 1963; Duchesneau 1974, 1982). "Penetrating forces" guide molecules in the development of forms, which thus continue to produce new life eternally; consequently, species also renew themselves. According to Buffon, the ontological core of each species is given, once and for all, by God, at the act of creation (Farber 1972).

Buffon's investigation was conjectural, based on analogy, which was quickly condemned by Sorbonne (1749–1750) and exerted a great influence on Diderot's materialist doctrine of life. In fact, two years after the publication of the first volume of the *Histoire naturelle*, Denis Diderot (1713–1784) wrote the article "Animal" for the *Encyclopédie* (Diderot 1751, Vol. I) and, in 1753, *Pensées sur l'interprétation de la nature* (thoughts on the interpretation of nature) (Diderot 1981). These two texts represent the first attempts at a philosophy of life that openly embrace Spinozist-inspired materialism and go far beyond the simple Lucretian model that still inspired Diderot in his *Letter on the Blind* (Diderot 1749[26]). The article "Animal" is a long quote from passages in the first volume of *Histoire naturelle*, interwoven with Diderot's philosophical considerations, which point toward an affirmation of the profound unity of the world of life. Two fundamental theses can be found:

– *nominalism*: "Species", as well as the classifications of naturalists (the "methodists," such as Linnaeus), are merely conventions, names assigned to categories of beings that are constantly changing, but from the perspective of being (of matter), nature only knows individuals (Leibniz *docet*[27]);

26 The *Letter* of 1749 adopts the ancient Epicurean doctrine of spontaneous generations (Diderot 1978; Quintili 2009a, 2009b).

27 "The more we examine nature, the more we become convinced that, in order to express ourselves accurately, we would need almost as many different names as there are individuals, and that it is necessity alone that has invented general names; given that these general names are more or less extensive, have meaning, or are meaningless, depending on how much progress we make in the study of nature" (Diderot 1751, *Encyclopédie*, I, p. 468b); on the importance of Leibniz, see Fauvergue (2006).

– *materialism*: natural individuals change and "pass away" on the basis of the ontological premise of the generalized sensitivity of matter, which ensures that every particle (Buffon's "organic molecule") forming organisms is equipped with this property and is therefore sensitive and "animated" (Spinoza *docet*).

To reach this conclusion, Diderot also studied the work of Maupertuis, who, before Buffon, had put forward bold theories on the structure of living beings. In 1745, the naturalist from Saint-Malo published a curious work named *La Vénus physique*, in which he explained the strange phenomena of the "albino negro". The particles of the father's and mother's seed (black), being equipped with sensitivity, memory, aversion, etc., in other words, with both psychic and material qualities in the case of a black child with little white skin, reproduce a conformation from the past, belonging to the ancestors of this new individual, among whom there *must have been a white parent* (Maupertuis 1745). This type of reasoning, which is based on both Leibnizian (living monads) and Newtonian theories – particles attract and repel one another according to the laws of *actio in distans* – is incorporated by Diderot in his *Interpretation of Nature*, in Thought 50. Here, another work by Maupertuis, which had just been published in 1751 under the pseudonym "Doctor Baumann", *Dissertatio de universali naturae systemate*, published in 1754 in French under the title *Essai sur la formation des corps organisés* (*Essay on the Formation of Organized Bodies*) (Maupertuis 1751, 1754), is discussed in greater detail. In this context, Maupertuis extends his conjecture of particles containing life, sensitivity and memory to the entire natural universe. Newton's laws of attraction, inherent in organic matter, can account for the processes of molecular aggregation, as well as the morphological laws of imitation, distinction, fetal development and, above all, *heredity*, which characterize the acts of living nature. However, Maupertuis recognizes here that the laws of attraction alone are not sufficient to explain the complexity of living beings[28].

To this end, by introducing a principle of "intelligence" into matter, Maupertuis not only paved the way for a monistic, non-Cartesian approach to phenomena concerning *res extensa*, which can be assimilated to the phenomenal order of *res cogitans*, but he also moved toward a continuous vision of the natural cosmos that challenged the old Neoplatonic and Leibnizian concept of the "scale of beings"

28 "A uniform and blind attraction spread throughout all parts of matter cannot explain how these parts arrange themselves to form the body, whose organization is the most basic. If they all have the same tendency, the same force to unite with one another, why do some form the eye, and others the ear? Why this marvelous arrangement? Why do they not all unite randomly? [...] We must resort to some principle of intelligence, to something similar to what we call *desire, aversion, memory*" (ibid., pp. 13–14).

(*scala naturae*[29]), modifying it from within, in its ontological meaning. The ontological order "on a scale" no longer implies an intrinsic *immutability* of living forms but also includes the principle that explains their differences, disruptions and deviations (the problem of "monsters" or "deviations from nature" (Roger 1963; Curran 2001)) in the organic whole that develops over time and through interaction with the natural world. Maupertuis states the following:

> In the fluid state in which matter was, each element will have positioned itself in the appropriate manner to form these bodies, in which no trace of their formation can be recognized. [...] thus, when a swarm of bees have gathered and united around the branch of some tree, our eyes perceive nothing more than a body that bears no resemblance to the individuals that formed it. (Ibid., pp. 47–48[30])

Faced with these questions, Diderot delved into speculation and, in *Interpretation*, reformulated the essential philosophical assumptions. Matter is one and sensitive; the dualism of substances – matter and spirit – cannot explain the profound unity of the phenomena of life; the gradual, "scaled" order of beings is not arranged on an immutable vertical scale of physical beings-man-God but is a *horizontal* and circular *flow* of interrelated mechanical-organic phenomena that gives rise to a "system" that is perfectly unified as a whole but contingent in its parts, in a state of fruitful and perpetual transformation.

Owing to Maupertuis' *Dissertatio* and Buffon's *Histoire naturelle*, Diderot developed a temporal and dynamic vision of the natural order, which could undoubtedly be called a "transformist"[31]. Why? The correlated elementary parts of the matter of organized bodies, with memory, desire and aversion, are assimilated by Diderot to organic molecules, which, according to Buffon, constituted the primordial nucleus of the development of the living natural universe. Buffon and Maupertuis' epigenesis is interpreted in a materialist way by Diderot, and the object of his criticism is the old thesis of the pre-existence of germs by Malebranche, Malpighi

29 On the ancient Neoplatonic (and Stoic) concept of the "chain of beings" in the modern age, see the classic work by Lovejoy (1936). Diderot, Maupertuis, Buffon and the Montpellier medical school marked a *temporalization* of the chain.

30 The (famous) metaphor of the swarm of bees is repeated in *Le Rêve de d'Alembert* (Diderot 1987, pp. 120–121).

31 On the controversial issue of "transformism" in the 18th century, the debate is still open among interpreters. Jacob (1970, p. 157) denies the presence of transformist thought in the 18th century before Lamarck; Quintili (2001, pp. 311–312) and Diderot (2004, pp. 60–63) argue in favor of its presence in Diderot's biological philosophy.

and Swammerdam, which he also considers in a conjectural manner to refute it through a comparison (Thought 32) with the theories of Charles Bonnet (1720–1793) (Bonnet 1745).

Diderot identifies the fundamental mechanical core of vital processes in the progressive development of contiguous/continuous parts. This results in a materialistic hylozoism or a general "animality" of the cosmos, which brings Maupertuis' system of intelligent particles closer, once again, to Spinoza's *Deus sive natura*[32]. The animal is then defined in *Interpretation*:

> A system of different organic molecules which, through the impulse of a sensation similar to a dull, blunt touch, such as that which created matter in general, combined until each found the place most suitable to its shape and position. (Diderot 1981, pp. 84–85)

Diderot (1769) reiterated these ideas, developing them on a large scale, in his posthumous masterpiece, *Le Rêve de d'Alembert* (*d'Alembert's Dream*) and, later, in *Éléments de physiologie* (*Elements of Physiology*). The emphasis in *Le Rêve* is on the question of the organic whole, of the "great whole" that organizes itself, and the only "great individual" in the universe, a materialist echo of Spinoza's substance, which is articulated in its "parts" or particles (modes) that depend on it (Diderot 1987, pp. 139–140 *et seq.*). Although it is an outcome or an emerging property of the very laws of this universally sensitive matter, its sensitivity can be "inert" (inorganic) or "active" (organic). The tendency toward "organization" is unique to living beings, and this is the "soul of the animal". Diderot pays close attention to the phenomena of the material formation of consciousness and identity ("the self") in animals. The brain is the organ of the soul, the "origin of the bundle" of sensitive fibers that organize themselves in an increasingly complex manner in the chain of beings to form the animal's self-awareness. In humans, it is *memory*, an organic faculty, which – like a "sensitive harpsichord", resonating and reproducing sounds,

32 "I will therefore ask him whether the universe, or the general collection of all sensitive and thinking molecules, forms a whole or not. If he answers that it does not form a whole, he will undermine the existence of God with a single word, introducing disorder into nature and destroying the basis of philosophy, breaking the chain that binds all beings together. If he agrees that it is a whole, in which the elements are no less ordered than the parts, or truly distinct, or are only intelligible in an element, and the elements in an animal; he will have to admit that, as a consequence of this universal copulation, the world, like a great animal, has a soul; that the world, being infinite, this soul of the world, I do not say is, but may be an infinite system of perceptions, and that the world may be God" (Diderot 1981, p. 82). On the debate between Diderot and Maupertuis, which continued after the publication of *Interpretation*, see Maupertuis (1984), Wolfe (2010) and Ibrahim (2019).

and "plucking itself" – ensures the continuity of states of consciousness beneath the flow of sensory perceptions of experience.

There is a very rich proliferation of these themes; after writing *Le Rêve* in 15 days in August 1769, Diderot decided to explore the themes further in a collection of notes and reflections written between 1769 and his death in 1784: *Éléments de physiologie* (*Elements of Physiology*) (Diderot 2004, pp. 11–101). This work, also posthumous, is a broad synopsis of Diderot's reading notes, which were inspired by *Elementa Physiologiae corporis humani* (von Haller 1759–1769), a work written by the Swiss naturalist Albrecht von Haller (1708–1777). The major achievements of this work, with respect to the question of vitalism, consist of:

– the definitive acquisition of a dynamic and temporal vision of the chain of beings, which extends further on the scale of "millions of centuries"[33];

– the understanding of the relationship between functions and needs, as the key to the morphological transformations of organisms[34];

– a complex vision of this temporal dynamic, with the doctrine of the "three lives": (a) the life of the molecule (eternal), (b) the life of the organ and (c) the life of the whole animal[35];

– the centrality of the brain as the *organ of the soul*, to which Diderot devotes Part III of his work.

In *Pensées sur l'interprétation de la nature* (*Thoughts on the Interpretation of Nature*), *Essai sur la formation des corps organisés* (*Essay on the Formation of Organized Bodies*), *Le Rêve de d'Alembert* (*D'Alembert's Dream*) and *Éléments de*

33 "The chain of beings is not interrupted by the diversity of forms. Form is often only a deceptive mask; and the missing link may reside in a known being, to whom the progress of comparative anatomy has not yet been able to assign its true place" (Diderot 2004, p. 108); "Let the present race of surviving animals pass away; let the great inert sediment act for a few million centuries. Perhaps ten times the time is needed to renew a species than is granted to their duration; wait and do not rush to pass judgement on the great work of nature" (Diderot 1987, p. 132).

34 "Animals are beings whose form is determined by internal and external causes, which, being diverse, must produce diverse animals. The organization of each determines its functions and needs, and sometimes needs influence organization" (ibid., p. 137).

35 "The brain, or the cerebellum with the nerves that are merely filamentous and solid extensions of it, form a sensitive, continuous, energetic, and living whole: we must not seek to understand how this whole lives, for we do not know. There are certainly two very distinct lives, even three. The life of the whole animal. The life of each of its organs. The life of the molecule. The whole animal lives stripped of many of its parts" (ibid., p. 126).

physiologie (*Elements of Physiology*), the philosophical foundations of *vitalism* from the Enlightenment era were established. The limitation of this research, which was conducted by philosophers (Roger 1963, Part III, pp. 585–682) – in addition to de La Mettrie, Maupertuis and Diderot, d'Holbach would also join their ranks – was purely hypothetical and conjectural in nature. There was no evidence or real experimental evidence to support the thesis of sensitive matter and particles with psychic properties. However, a large group of scientists and physicians, both engaged in practice and research, looked into the same questions and provided new arguments for the vitalist approach on the basis of experience. These were the physicians of the Montpellier School who collaborated with Diderot in writing *Encyclopédie*, which were Théophile de Bordeu (1722–1776), Henri Fouquet (1727–1806), Jean-Jacques Ménuret de Chambaud (1739–1815) and Paul-Joseph Barthez (1734–1805[36]).

De Bordeu had been the protagonist of d'Alembert's *Le Rêve*, the character who conversed with Miss de L'Espinasse, d'Alembert's lover, on the arguments of the genesis of life, the universe populated by life, consciousness and the self (with a scandalous "appendix" on the lawfulness of masturbation and the mixing of species). However, the Béarnese physician wrote only one important article for *Encyclopédie*, but it was very long: "Crise" (de Bordeu 1751, *Enc.* IV, pp. 471b–489b). He contested the use of quantitative "mechanical reasoning" in favor of a "sort of philosophical or transcendent medicine, which perhaps all popular physicians, I mean clinical physicians, should adhere to" (de Bordeu 1751, *Enc.* IV, p. 489a). The crises of diseases are explained by a complex interplay of organs, each of which has its own time and specific action of excretion and secretion. Diderot had already been conscious of the arguments that de Bordeu had put forward in his masterpiece, *Recherches anatomiques sur la position des glandes et sur leur action* (*Anatomical Research on the Position of Glands and Their Action*)[37], in which he had demonstrated this "philosophical medicine" in the form of an organicist physiology and an "organology" which, following the teachings of Maupertuis and Spinoza, made each part of the living body an *animal in animali*, a separate living being that contributes to the life and functioning of the whole, that is to say, of the entire "machine". Importantly, even from this vitalist perspective, the metaphors

36 On the whole question of vitalism in *Encyclopédie*, see Rey (2000, pp. 59–71).

37 "Secretion is therefore reduced to a kind of *sensation*, if one may express it thus; the parts capable of exciting such a sensation will pass, and the others will be rejected; each gland, each orifice will, so to speak, have its own particular taste; anything foreign to it will normally be rejected [...], thus an animal is composed of different parts, each belonging to different kingdoms of Nature" (de Bordeu 1751, pp. 373–374) (see Quintili (2009a, 2009b, pp. 91–110); Mortier-Trousson (1999, pp. 83–85)).

used to model living beings are *mechanical* in nature, drawing on the image of a machine, a vitalist micromechanism that goes far beyond the old mechanistic view[38]. The living body is a genuine "association" of organs, supported by the "tripod of life": the brain, the heart and the diaphragm, which are the primary centers of vital force. With his study on the functioning of the glands, de Bordeu, who is sometimes considered the precursor (or pioneer) of endocrinology (Kafker 1988, pp. 44–47), even imagined the possibility of transplanting organs, each having a life that was independent from that of the whole, to a certain extent[39].

Fouquet's contribution to *Encyclopédie* is much more significant, consisting of six remarkable articles from the perspective of Montpellier vitalism: "Plomb" (Lead), "Sensitivity, Sentiment (*Medicine*)", "Secretion", "Burn", "Cupping" and "Vesicant". The second, in particular (Fouquet 1751, *Enc.* XV, pp. 38b–52b), is a true vitalist treatise written across 15 folio pages on the biological concept of "sensitivity" of the body and organs, inspired by the theories of his master and mentor de Bordeu (Dulieu 1952). The article first presents a general definition of bodily or organic sensitivity:

> The ability to feel, the sensitive principle, or the very feeling of the parts, the basis and preserving agent of life, animality at its finest, the most beautiful, the most singular phenomenon of nature. Sensitivity is in the living body, a property that certain parts have of perceiving impressions from external objects and of producing movements proportionate to the degree of intensity of this perception. (Fouquet 1751, *Enc.* XV, p. 38b)

This "property", according to Fouquet, governs the actions of the body. It is in action through its organs, each of which has its own sensitivity, and it is internal to matter or independent of any other "force" that is external to the organism[40]. It is a

38 "In order to understand the particular action of each part, we compare the living body to a swarm of bees that gather in clusters and hang from a tree like a bunch of grapes; the Elder's description of the viscera of the lower abdomen as an *animal in animali* was not incorrect; each part is, so to speak, not an animal, but a kind of separate machine, which contributes in its own way to the *general life* of the body" (ibid., p. 452). The same image of a cluster of bees appears in Diderot (1987) and Maupertuis (1754).

39 On this issue as a whole, see Dulieu (1983) and Quintili (2001, Chap. VI, pp. 295–353).

40 "From all that we have just said, it follows that we can regard feeling in animals as a physical passion or a property of matter, without needing to resort to the spiritual soul that judges or assesses sensations, as Stahl claims, in order to explain the terrible spasms that even a slight stimulus can cause" (Fouquet 1751, *Enc.* XV, p. 40a).

small step from here to physio-psychological materialism, even if Fouquet retains the Aristotelian distinction between the "sensitive soul" and the "rational soul" and makes the former a kind of substance of life, as Barthez does with the "vital principle". In his *Elements of Physiology*, Diderot did not hesitate to take this step. Fouquet, who had discussed his thesis in Montpellier in 1759 in *De fibra natura viribus et morbis in corpore animali*, then took sides on the question that had opposed de Bordeu to Glisson (1597–1677) (Glisson 1672) and von Haller (1753), who identified a new physical property in the animal body that would be the basis of phenomena involving organ tissues and fibers: *irritability* (or contractility), according to von Haller, as opposed to sensitivity, which is specific to nerves. Fouquet supported the unique, *living* nature of the fiber as the primary element of the organism, the source of sensitivity and movement even in the formation of the fetus[41]. As Diderot would say, "In physiology, the fiber is what the line is in mathematics" (Diderot 2004, p. 156).

Ménuret de Chambaud, the youngest member of the "gang" of vitalists, was the most prolific writer, having written 72 signed articles and having 16 attributed to him, starting with Volume VIII of *Encyclopédie* (1765). It is difficult to summarize his doctrine, which, as Rey (2000) noted, represents the following:

> The convergence of various doctrines [...] the rejection of systems, eclecticism, and the importance of personal medical observations make Ménuret an exemplary representative of the Montpellier School. (Rey 2000, p. 61)

He thus contributed to spreading the vitalist theories of his school on fundamental points, especially in his major articles "Inflammation, Inflammatory Disease", "Observation" (and reasoned sequence), "Animal Economy", "Pulse" and "Death". The article "Inflammation" is a treatise on this new form of medicine that controversially attacks the "mechanists" – the dominant feature of Ménuret's style is controversy – on one fundamental point:

> The mistakes made by Mechanists seem, to me, to stem from no other principle than that they have not considered animals as composite,

41 "This cylinder [the rudiments of the fetus], which could either be called *animal fiber* or *animal atom*, endowed with the spiritual soul in man, grows more and more, appropriating the molecules of the fluid that surrounds it; it becomes covered with asperities and sprouts small branches in all directions, tracing the delineations of the parts, in accordance with the pattern imprinted by the Creator. Ultimately, all the organs develop under the activity of the offshoots of this first and only nerve, which work in different ways on the highly *ductile mucus* to construct themselves, like houses" (Fouquet 1751, *Enc.* XV, p. 40b).

> living, and organized beings. However, even if we were forced to admit a motor faculty that acts and operates in the body, it would have to be considered different from the soul and intended to regulate vital movements. (de Chambaud 1751, *Enc.* VIII, p. 713b)

This vital or motor function is material, such as the irritability/sensibility described by de Bordeu and Glisson, and acts equally in humans and all animals. It explains all forms of morbidity, "fever, inflammation, and other phenomena of animal economy that were previously attributed to the soul or nature" (ibid.). This thesis is reiterated in the articles "Pulse" – nearly 50 pages, from which the author drew his *Nouveau traité du pouls* (New Treatise on the Pulse) (de Chambaud 1769) – "Observation" and "Animal Economy", his masterpiece. Here, Ménuret announces the "great revolution" introduced by the new "philosophical spirit" of the century, which consists of the renewal of the criteria of observation and the interpretation of medical facts, according to the principle of the vital and/or motor function of every organism[42]. The points of reference are Buffon, Diderot and de Bordeu, whom Ménuret considers geniuses of reasoned reflection and observation. According to the article "Observation", iatromechanics have neglected accurate observation and, above all, clinical practice, with their "sterile taste for constantly fruitless experiments" (de Chambaud 1751, *Enc.* XI, p. 316a). In reality, the animal economy shows us that the physician must grasp everything and see the whole organism of the animal to be able to follow the morbid development of its separate parts. Ménuret is the most "holistic" vitalist among the Montpellier school. The "animal economy" – "the order, mechanism, and set of functions and movements that sustain the life of animals" – already demonstrates this primacy of the whole in its etymology: the common "house" (*oikòs*), whose laws (*nòmoi*) must be found, is the whole animal:

> The human body is a *statico-hydraulic* type of machine, composed of solids and fluids, whose primary elements, shared by plants and animals, are *living atoms*, or *organic molecules*: let us imagine the marvelous assembly of these molecules, as anatomical observations reveal them to us [...] this machine thus formed only differs from the

42 "The philosophical spirit that is entering medicine, which mainly seeks facts, which leads one to see everything, to examine everything, to grasp the truth with eagerness and to love it above all else; the enormous number of past errors, which leaves us with less to fear, and perhaps also the enlightenment of our age, all these causes combined favor the return of *observation* and serve to rekindle this torch. Medicine seems to be on the verge of a great revolution; well-regarded systems are being reduced to their true value; many doctors are applying themselves properly to *observation*; they are following nature, and they will not delay in reviving the medicine of Hippocrates, which is the true medicine of *observation*" (de Chambaud 1751, *Enc.* XI, p. 316b).

> living human being in terms of movement and sensation, the principal phenomena of life, which are probably reducible to a single primitive phenomenon (de Chambaud 1751, *Enc.* XI, p. 361a), which is irritability/sensibility.

De Bordeu and Diderot assert that "life is nothing but movement and feeling". On this point, the School is unanimous. Pierre-Joseph Barthez takes this general dual vitalist doctrine "principle" to its conclusion, unifying it and giving it the proper term "vital principle", or principle of life[43]. Barthez rewrote 22 articles for *Encyclopédie*, in volumes VI–VII (Barthez 1756–1757), which are not among the most important or original in the work: "Fainting (*Medicine*)", "Expansion (*Anatomy*)", "Extensor (*Anatomy*)", "External (*Anatomy*)", "Face (*Anatomy*)", "Faulx (*Anatomy*)", etc. Most of these articles are on anatomy and show almost no trace of his colleagues' vitalist theses, even on relevant topics. Barthez's sources are often outdated (Galen, Albinus, Borelli, etc.) even in the longer articles, "Woman" (*Anthropology*) and "Force of Animals". At a very young age (22), the physician limited himself to compilations. "Fibre (*Anatomy*)" and "Fibrille (*Anatomy*)", for example, do not reflect the theories of Glisson and von Haller on irritability, which had captivated de Bordeu, Fouquet and Ménuret. Barthez's contribution to the vitalist school lies elsewhere, notably in the works published after 1765: *Oratio academica de principio vitali hominis* (Barthez 1772), *Nouveaux éléments de la science de l'homme*[44] (Barthez 1778) and *Nouvelle mécanique des mouvements progressifs de l'homme et des animaux*[45] (Barthez 1798). De Bordeu (1767), however, the former master, quickly grasped and recorded Barthez's new use of an old notion, also used by iatromechanics, the "vital principle" and emphasized its problematic nature in his *Recherches sur les maladies chroniques*[46]:

> The name *vital principle* was beginning to grow old; but it has just been given new life by one of Fizes' successors. Mr. Barthés [sic], rising well above his predecessor but retaining only his expression. He is not a mechanic like Fizes; but he follows him in his distaste for the *nature of the Elders*, for the *arche*, for the *soul* of the Stahlians, and perhaps for *sensibility and vital mobility*. Thus, the *vital principle* is no longer the mechanics of the body dependent on its structure: it is

43 "Diseases are essentially the result of disorders of the Principle of Life in man; [...] diseases are generally determined automatically by the action of morbid causes, whether they be external or internal, in accordance with laws established by the Vital Principle, which are neither mechanical nor arbitrary" (Barthez 1778, p. 43, cited in Rey (2000, p. 236)).

44 *New Elements of Human Science* (Barthez 1778).

45 *New Mechanics of Progressive Movements in Humans and Animals* (Barthez 1798).

46 Research on chronic diseases (de Bordeu 1767).

> not nature, it is not the soul, it is not the sensibility of the animal element: how and in what way does it differ? It is Mr. Lamure and Mr. Venel, and then Mr. Fouquet, who openly declared himself in favor of sensibility, who will clarify on this issue. (de Bordeu 1767, p. 366)

This is an important page because this "principle" that Barthez would adopt to explain the vital phenomena of the organism would take on a metaphysical connotation – as a form of substance or a "kind of pre-established harmony between the affections of this principle and the organization of the body" (Barthez 1778, Vol. II, p. 202) – which it had not previously had and which the encyclopedists would not accept, including Diderot himself (Diderot 2004, p. 126 and *Introduction*, pp. 42–43). With respect to the specific explanation of "vapors", Barthez distinguished himself by his approach, which seemingly circumvented the question of the nature of the vital principle, to focus on the action of the organic driving forces of "sympathy", as opposed to "sensory forces" – feeling + movement = life – which, when disrupted, cause morbid affection[47]. The legacy of the vitalist school's encyclopedic lesson, through Barthez, in the second half of the 18th century, thus expressed all the ambiguities and hesitations – embodied in this "unknown" that is the "vital principle"[48] – of a multifaceted doctrine that nevertheless would have a profound impact on the scientific and philosophical culture of the centuries to come (Mitchell 2012; Packham 2012; Normandin and Wolfe 2013; Quintili 2009b).

7.5. References

Primary sources

Aristotle (2014). *Œuvres complètes*, Pellegrin, P. (ed.). Flammarion, Paris.

Barthez, P.-J. (1772). *Oratio academica de principio vitali hominis*. A.F. Rochard, Montpellier.

47 "When this general change in the vital forces exists to a very high degree, without any permanent lesion of any particular organ as its main cause, it constitutes the disease that has been given the name of vapors, and which has also been called neuropathy [...] The main symptoms of vapors clearly depend on an exaltation or decrease of the natural activity of the sensitive forces; and from a depravation of their influence on the motor forces" (Barthez 1778, Vol. II, pp. 174–175; Rey 2000, pp. 238–239).

48 "Although, for greater convenience, the Vital Principle is constantly personified in this work, it can be replaced by the abstract notion of a simple vital faculty that is *unknown* in its essence. – The Unity of the Vital Principle is established by various facts" (ibid., Vol. II, p. 201).

Barthez, P.-J. (1778). *Nouveaux éléments de la science de l'homme*, volume 2. J. Martel, Montpellier.

Barthez, P.-J. (1798). *Nouvelle mécanique des mouvements progressifs de l'homme et des animaux*. Méquignon l'ainé, Paris.

Bayle, P. (1697). *Dictionnaire historique et critique*, volume 4. R. Leers, Rotterdam.

Boerhaave, H. (1739). *Institutions de médecine*, volume 1, J.O. de La Mettrie (trans.). Huart/Briasson, Paris.

Boerhaave, H. (1740). *Institutions de médecine*, volume 2, J.O. de La Mettrie (trans.). Huart/Briasson, Paris.

Bonnet, C. (1745). *Traité d'insectologie ou observations sur quelques espèces de vers d'eau douce, qui coupés par morceaux deviennent autant d'animaux complets*. Chez Durand, Paris.

de Bordeu, T. (1751). *Recherches anatomiques sur la position des glandes et sur leur action*. G.F. Quillau, Paris.

de Bordeu, T. (1767). *Analyse médicale du sang*, in *Recherches sur les maladies chroniques, leurs rapports avec les Maladies aigues, leurs périodes, leur nature*. T.I. Ruault, Paris.

Borelli, G.A. (1680). *De motu animalium pars prima*. Ex Typographia Angeli Bernabò, Rome.

Borelli, G.A. (1681). *De motu animalium pars altera*. Ex Typographia Angeli Bernabò, Rome.

de Buffon, G.-L.L. (2007a). Histoire naturelle générale et particulière. Tome I : Histoire et théorie de la Terre. In *Œuvres*, Schmitt, S. (ed.). Gallimard Pléiade, Paris [Original published 1749].

de Buffon, G.-L.L. (2007b). Histoire naturelle générale et particulière. Tome II : Histoire des animaux. De la nature de l'Homme. In *Œuvres*, Schmitt, S. (ed.). Gallimard Pléiade, Paris [Original published 1749].

de Condillac, E.B. (1755). *Traité des animaux, où, après avoir fait des observations critiques sur le sentiment de René Descartes, et sur celui de M. de Buffon, on entreprend d'expliquer leurs principales facultés*. De Bure-Jombert, Amsterdam/Paris.

Descartes, R. (1949). *Œuvres et lettres*, Bridoux, A. (ed.). Gallimard, Paris.

Descartes, R. (1996). L'Homme. In *Œuvres*, volume XI, Adam, C. and Tannery, P. (eds). Vrin, Paris [Original published 1662/1664].

Descartes, R. (2000). *Écrits physiologiques et médicaux*, Aucante, V. (ed.). PUF, Paris.

Diderot, D. (ed.) (1751–1772). *Encyclopédie, ou Dictionnaire raisonné des sciences, des arts et des métiers*, 17 volumes. Le Breton/Briasson/Durand, Paris.

Diderot, D. (1978). Lettre sur les aveugles. In *Œuvres complètes, volume IV Le nouveau Socrate*, Dieckmann, H., Proust, J., Varloot J. (eds). Hermann, Paris [Original published 1749].

Diderot, D. (1981). Pensée sur l'interprétation de la nature. In *Œuvres complètes, volume IX Idées III*. Hermann, Paris.

Diderot, D. (1987). Le Rêve de D'Alembert. In *Œuvres complètes, volume IX Idées IV*. Hermann, Paris [Original published 1769].

Diderot, D. (2004). *Éléments de physiologie, en Appendice le Manuscrit de Pétersbourg*, Quintili, P. (ed.). Honoré Champion, Paris.

Galien, C. (1994). *Œuvres médicales choisies*, Daremberg, C. (trans.), Pichot, A. (ed.). Gallimard Tel, Paris.

Galilei, G. (1632). *Dialogo di Galileo Galilei Linceo, dove ne' congressi di quattro giornate si discorre sopra i due massimi sistemi del mondo, tolemaico e copernicano*. Gio. Battista Landini, Firenze.

Galilei, G. (1691–1692). Dialogo de massimi sistemi del mondo tolemaico e copernicano. In *Le Opere di Galileo Galilei, nobile fiorentino*, volume VII, Giacopo Monti, Bologna.

Galilei, G. (1972). *Discours et démontrations mathématiques concernant deux sciences nouvelles*, Clavelin, M. (ed.). PUF, Paris [Original published 1638].

Galilei, G. (1992). *Dialogue sur les deux grands systèmes du monde*, Fréreux, R. (trans.), De Gandt, F. (ed.). Le Seuil, Paris [Original published 1632].

Gassendi, P. (1658). Syntagma Philosophicum. In *Petri Gassendi Diniensis Ecclesiæ Præpositi et in Academia Parisiensi Matheseos regii professoris Opera Omnia*, 6 volumes. Laurent Anisson, Lyon.

Glisson, F. (1672). *Tractatus de natura substantiae energetica, seu de vita naturae ejusque tribus primis facultatibus I. Perceptiva, II. Appetitiva, et III. Motiva*. Brome, London.

de Graaf, R. (1668). *De virorum organis generationi inservientibus, de clysteribus et de usu siphonis in anatomia*. Ex Officina Hackiana, Leiden and Rotterdam.

von Haller, A. (1753). De partibus corporis humani sensibilibus et irritabilibus. In *Commentarii Societatis regiæ scientiarum gottingensis, ad annum 1752*, volume II, Göttingen.

von Haller, A. (1757–1769). *Elementa physiologiae corporis humani*, 8 volumes. Bousquet, Lausanne.

Hartsoeker, N. (1706). *Conjectures physiques*. H. Desbordes, Amsterdam.

Hartsoeker, N. (1710). *Éclaircissements sur les Conjectures physiques*. P. Humbert, Amsterdam.

Hartsoeker, N. (1712). *Suite des Éclaircissements sur les Conjectures physiques*. N. Viollet, Amsterdam.

Hartsoeker, N. (1717). Lettre sur quelques endroits des Ouvrages de Mrs Cheyne et Derham, à l'auteur de la Bibliothèque ancienne et moderne. In *Bibliothèque Ancienne et Moderne. Pour servir de suite aux Bibliothèques Universelle et Choisie*, volume VII, Le Clerc, J. (ed.). David Mortier, Amsterdam.

Hartsoeker, N. (1730). *Cours de Physique*. J. Swart, La Haye.

Harvey, W. (1628). *Exercitatio anatomica de motu cordis et sanguinis in animalibus*. Sumptibus Guilielmi Fitzeri, Frankfurt.

van Helmont, J.-B. (1648). *Ortus medicinæ, id est initia physicæ inaudita. Progressus medicinæ novus*. Louis Elzevier, Amsterdam.

de La Mettrie, J.O. (1739). *Nouveau traité des maladies vénériennes : par M. de La Mettrie, docteur en médecine*. Huart, Paris.

de La Mettrie, J.O. (1748). *L'homme machine*. Elie Luzac, Leiden.

de La Mettrie, J.O. (1987 [1750]). Les animaux plus que machines. In *Œuvres philosophiques*, volume II, Markovitz, F. (ed.). Fayard, Paris.

van Leeuwenhoek, A. (1687). *Anatomia seu interiora rerum, cum animatarum tum inanimatarum, ope et beneficio exquisitissimorum microscopiorum detecta*. Boutesteyn, Lugduni Batavorum.

van Leeuwenhoek, A. (1689). *Continuatio epistolarum datarum ad longe celeberrimam Regiam Societatem Londinensem*. Boutesteyn, Lugduni Batavorum.

Leibniz, G.W. (1768). *Opera omnia*, 6 volumes. Les frères de Tournes, Geneva.

Leibniz, G.W. (1890). Commentatio de anima brutorum. In *Die philosophischen Schriften von G. W. Leibniz*, volume VII, Gerhardt, C.J. (ed.). Berlin [Original published 1710].

Leibniz, G.W. (2016). *The Leibniz-Stahl Controversy*, F. Duchesneau, J.E.H. Smith, (trans.). Yale University Press, New Haven.

de Maillet, B. (1748). *Telliamed, ou Entretiens d'un philosophe indien avec un missionnaire français, sur la diminution de la mer, la formation de la Terre, l'origine de l'Homme, &c. Mis en ordre sur les Mémoires de feu M. de M*** par J. A. Guer. Avocat.*, 2 volumes. Chez l'honoré & Fils Libraires, Amsterdam/La Haye.

Malebranche, N. (1674). *De la recherche de la vérité. Où l'on traite de la nature de l'esprit de l'homme, et de l'usage qu'il en doit faire pour éviter l'erreur dans les sciences*. André Pralard, Paris.

Malpighi, M. (1673). *Dissertatio epistolica de formatione pulli in ovo*. Apud Joannem Martyn, London.

de Maupertuis, P.-L.M. (1745). *Vénus physique*. Paris.

de Maupertuis, P.-L.M. (1751). *Dissertatio de universali naturae systemate*. Erlangen.

de Maupertuis, P.-L.M. (1754). *Essai sur la formation des corps organisés*. Berlin.

de Maupertuis, P.-L.M. (1984). *Essai de cosmologie. Système de la nature. Réponse aux objections de M. Diderot*, Azouvi, F. (ed.). Vrin, Paris.

Ménuret de Chambaud, J.-J. (1769). *Nouveau traité du pouls*, 2 volumes. P.-F. Didot le Jeune, Paris.

Perrault, C. (1680). *Essais de physique*. J.-B. Coignard, Paris.

Rorarius, N. (1566). *Contradictiones, dubia, et paradoxa in Libros Hippocratis, Celsi, Galeni, Aetii, Aeginetae, Avicennae. Cum eorundem Conciliationibus*. Apud Franciscum/ Gasparem Bindonum, Venice.

Spinoza, B. (1677). *Ethica, ordine geometrico demonstrata*. Jan Rieuwertsz, Amsterdam.

Spinoza, B. (2020). *Éthique. Texte latin établi par C. Appuhn ; traduction, introduction et notes par C. Appuhn. Nouvelle édition revue*. Presses Universitaires de France, Paris.

Stahl, G.-E. (1708). *Theoria medica vera, physiologiam et pathologiam, tanquam doctrinae medicae partes vere contemplativas, e naturae et artis veris fundamentis, intaminata ratione et inconcussa experientia sistens*. Literis Orphanotrophei, Halle.

Stahl, G.-E. (1766). *Traité du soufre, ou remarques sur la dispute qui s'est élevée entre les chymistes, au sujet du soufre, tant commun, combustible ou volatil, que fixe &c.*, D'Holbach (trans.). P.-F. Didot, Paris.

Stahl, G.-E. (1771). *Traité des sels, dans lequel on démontre qu'ils sont composés d'une terre subtile, intimement combinée avec de l'eau, par George-Ernest Stahl*, D'Holbach (trans.). Vincent, Paris.

Stahl, G.-E. (1860). Paraenesis ad aliena a medica doctrina arcendum. In *Œuvres médico-philosophiques et pratiques*, 6 volumes, Blondin, T. (ed.). Bailliere, Paris .

Swammerdam, J. (1669). *Historia Insectorum generalis, ofte Algemeene Verhandelung van de Bloedeloose Dierkens*. M. van Dreunen, Utrecht.

Vesalius, A. (1555). *De humani corporis fabrica*. Per Ioannem Oporinum, Basel.

Willis, T. (1672). *De Anima Brutorum quæ Hominis Vitalis ac Sensitiva est, Exercitationes duæ. Prior PHYSIOLOGICA Ejusdem Naturam, Partes, Potentias & Affectiones tradit. Altera PATHOLOGICA Morbos qui ipsam, & sedem ejus Primariam, nempe Cerebrum & Nervorum Genus afficiunt, explicat, eorumque Therapeias instituit*. E Theatro Sheldoniano, Impensis Ric. Davis, Oxford.

Secondary sources

Almog, J. (2002). *What Am I? Descartes and the Mind-Body Problem*. Oxford University Press, Oxford.

Andrault, R. (2014). *La vie selon la raison. Physiologie et métaphysique chez Spinoza et Leibniz*. Honoré Champion, Paris.

Andrault, R. (2016). *La raison des corps. Mécanisme et sciences médicales (1664-1720)*. Vrin, Paris.

Andrault, R. (2018). O que pode o corpo ? Spinoza, na cabeceira dos esfolados. *O Que nos faz pensar*, 26(41) [Online]. Available at: http://www.oquenosfazpensar.fil.puc-rio.br/index.php/oqnfp/article/view/583.

Aucante, V. (2006). *La philosophie médicale de Descartes*. PUF, Paris.

Barcaro, U. (1982). Galileo e il mondo biologico. *Physis*, 24, 357–379.

Benítez, M. (1980). Benoît de Maillet et la littérature clandestine. Étude de sa correspondance avec l'abbé Le Mascrier. *Studies on Voltaire and the Eighteenth Century*, 183, 133–159.

Bloch, O.R. (1982). *Le matérialisme du XVIII*e *siècle et la littérature clandestine*. Actes/Vrin, Paris.

Bloch, O.R. (1990). *Spinoza au XVIII*e *siècle*. Actes/Klincksieck, Paris.

Bouveresse, R. (1992). *Spinoza et Leibniz. L'idée d'animisme universel. Étude suivie de la traduction inédite d'un texte de Leibniz sur l'Éthique de Spinoza et d'un texte de Louis Meyer*. Vrin, Paris.

Canguilhem, G. (1971). *La connaissance de la vie*. Vrin, Paris.

Canguilhem, G. (1977a). *La formation du concept de réflexe aux XVII*e *et XVIII*e *siècles*. Vrin, Paris.

Canguilhem, G. (1977b). *Idéologie et rationalité dans l'histoire des sciences de la vie*. Vrin, Paris.

Canguilhem, G. (2002). *Études d'histoire et de philosophie des sciences concernant le vivant et la vie*. Vrin, Paris.

Carter, R.B. (1983). *Descartes' Medical Philosophy. The Organic Solution to the Mind-Body Problem*. Johns Hopkins University Press, Baltimore/London.

Castiglioni, A. (1942). Galileo Galilei and his influence on the evolution of medical thought. *Bulletin of the History of Medicine*, 12, 226–241.

Cohen Rosenfield, L. (1940). *From "Beast-Machine" to "Man-Machine". Animal Soul in French Letters from Descartes to La Mettrie*. Oxford University Press, New York.

Comte-Sponville, A. (1990). *La Mettrie : un 'Spinoza moderne' ?* O. Bloch, Paris.

Cosmacini, G. (1997). *L'arte lunga. Storia della medicina dall'Antichità a oggi*. Laterza, Rome/Bari.

Crane, T. and Patterson, S. (2000). *History of the Mind-Body Problem*. Routledge, London.

Croft, W.J. (2006). *Under the Microscope. A Brief History of Microscopy*. World Scientific Publishing Company, London.

Curran, A. (2001). *Sublime Disorder. Physical Monstrosity in Diderot's Universe*. Voltaire Foundation, Oxford.

Debus, A.G. (1977). *The Chemical Philosophy. Paracelsian Science and Medicine in the Sixteenth and Seventeenth Centuries*, 2 volumes. Science History Publications, New York.

Debus, A.G. (2014). *La médecine chimique dans le premier âge moderne*. M.D. Grmek, Paris.

Duchesneau, F. (1974). Du modèle cartésien au modèle spinoziste de l'être vivant. *Canadian Journal of Philosophy*, 3(4), 539–562.

Duchesneau, F. (1982). *La physiologie des lumières. Empirisme, modèles et théories*. Martinus Nijhoff, La Haye/Boston/London.

Duchesneau, F. (1998). *Les modèles du vivant de Descartes à Leibniz*. Vrin, Paris.

Dulieu, L. (1952). Les articles d'Henri Fouquet dans l'Encyclopédie. *Revue d'histoire des sciences et de leurs applications*, 5(1), 18–25.

Dulieu, L. (1983). *La Médecine à Montpellier. Volume III : l'Époque classique*. Les Presses Universelles, Avignon.

Farber, P.L. (1972). Buffon and the concept of species. *Journal of the History of Biology*, 5(2), 259–284.

Fauvergue, C. (2006). *Diderot lecteur et interprète de Leibniz*. Honoré Champion, Paris.

Gregory, T. (1955). *Anima mundi : la filosofia di Guglielmo di Conches e la scuola di Chartres*. G.C. Sansoni, Florence.

Gregory, T. (2000). *Genèse de la raison classique de Charron à Descartes*, M. Raiola (trans.). PUF, Paris.

Grmek, M.D. (2014a). *Histoire de la pensée médicale en Occident. Tome 1 : de l'Antiquité au Moyen-Age*, 2nd edition. Le Seuil, Paris.

Grmek, M.D. (2014b). *Histoire de la pensée médicale en Occident. Tome 2 : de la Renaissance aux Lumières*, 2nd edition. Le Seuil, Paris.

Hazard, P. (1938). *La crise de conscience européenne. 1680-1715*. Boivin, Paris.

Hoffman, P. (1996). *Théories et modèles de la liberté au XVIII*[e] *siècle*. PUF, Paris.

Ibrahim, A. (1990). *Sur le spinozisme dans les philosophies du vivant*. O. Bloch, Paris.

Ibrahim, A. (2019). Matérialisme et métaphysique : Diderot, Maupertuis, Dom Deschamps. In *Bulletin de la Société Française de Philosophie*, volume 3. Vrin, Paris.

Israel, J.I. (2001). *Radical Enlightenment. Philosophy and the Making of Modernity. 1650-1759*, P. Hugues, C. Nordmann, J. Rosanvallon, (trans.). Éditions Amsterdam, Paris/Oxford.

Jacob, F. (1970). *La logique du vivant. Une histoire de l'hérédité*. Gallimard, Paris.

Kafker, F. (1988). *The Encyclopedists as Individuals. A Biographical Dictionary of the Authors of the Encyclopédie*. Voltaire Foundation, Oxford.

King, L.S. (1958). *The Medical World of the Eighteenth Century*. University of Chicago Press, Chicago.

King, L.S. (1964). Stahl and Hoffman. A study in 18th century animism. *Journal of the History of Medicine and Allied Sciences*, 19, 118–130.

King, L.S. (1978). *The Philosophy of Medicine. The Early Eighteenth Century*. Cambridge University Press, Cambridge.

Knoeff, R. (2002). *Hermann Boerhaave (1668-1638). Calvinist Chemist and Physician*. Royal Netherlands Academy of Arts and Sciences, Amsterdam.

Koyré, A. (1937). Galilée et Descartes. In *Travaux du XIe Congrès international de Philosophie (Congrès Descartes)*, volume II. Hermann, Paris.

Lagerlund, H. (2007). *Forming the Mind. Essays on the Internal Senses and the Mind/Body Problem from Avicenna to the Medical Enlightenment*. Springer, Dordrecht.

Latronico, N. (1955). *Il cuore nella storia della medicina*. A. Recordati, Milan.

Lodoli, F. and Quintili, P. (2020). Spinoza nella letteratura clandestina francese. In *La fortuna di Spinoza in età moderna e contemporanea*, volume 1, Altini, C. (ed.). Seicento e Settecento/Edizioni della Normale, Pisa.

Lovejoy, A.O. (1936). *The Great Chain of Being. A Study of the History of an Idea*. Harvard University Press, Cambridge/London.

Macherey, P. (1997). *Introduction à l'Éthique de Spinoza. La seconde partie, la réalité mentale*. PUF, Paris.

Marcialis, G. (1993). La querelle sur l'âme des bêtes au XVIIe siècle. *Rivista di storia della filosofia*, 48, 215–240.

Mitchell, R. (2012). *Experimental Life. Vitalism in Romantic Science and Literature*. Johns Hopkins University Press, Baltimore.

Moreau, P.-F. (1994). *Spinoza. L'expérience et l'éternité*. PUF, Paris.

Moreau, P.-F. (2003). *Spinoza et le spinozisme*. PUF, Paris.

Mortier, R. and Trousson, R. (1999). *Dictionnaire de Diderot*. Honoré Champion, Paris.

Normandin, S. and Wolfe, C.T. (2013). *Vitalism and the Scientific Image in Post- Enlightenment Life Science, 1800–2010*. Springer, Dordrecht/Heidelberg/New York/London.

Oldroyd, D. (1973). An examination of G. E. Stahl's 'principles of universal chemistry'. *Ambix*, 20, 36–52.

Packham, C. (2012). *Eighteenth-Century Vitalism. Bodies, Culture, Politics*. Palgrave Macmillan, New York.

Pagel, W. (1984). *Paracelsus: An Introduction to Philosophical Medicine in the Era of the Renaissance*. Karger, New York.

Popper, K.R. (2002). *The Logic of Scientific Discovery*. Routledge, London [Original published 1935].

Quintili, P. (2001). *La pensée critique de Diderot. Matérialisme, science et poésie à l'âge de l'Encyclopédie. 1742-1782*. Honoré Champion, Paris.

Quintili, P. (2009a). *Anima, mente e cervello : alle origini del problema mente-corpo, da Descartes all'Ottocento*. Unicopli, Milan.

Quintili, P. (2009b). *Matérialismes et Lumières. Philosophies de la vie, autour de Diderot et de quelques autres. 1706-1789*. Honoré Champion, Paris.

Quintili, P. (2013). Entre Spinoza et Diderot. Le seuil du dévoilement de l'athéisme. In *Athéisme dévoilé aux temps modernes*, Staquet, A. (ed.). Éditions de l'Académie, Brussels.

Quintili, P. (2024). *Mind-Body Problem. Mente, corpo, emozioni e passioni*. Mimesis, Milan.

Rey, R. (2000). L'Encyclopédie et la science des animaux. In *Lire l'Encyclopédie. Le "Système figuré" des connaissances humaines*, Rey, A. (ed.). Gallimard, Paris.

Roger, J. (1963). *Les sciences de la vie dans la pensée française du XVIIIe siècle. La génération des animaux de Descartes à l'Encyclopédie*. Armand Colin, Paris.

Singer, C. (1959). *A Short History of Scientific Ideas to 1900*. Oxford University Press, London.

Solinas, G. (1967). *Il microscopio e le metafisiche. Epigenesi e preformismo da Cartesio a Kant*. Feltrinelli, Milan.

Thomson, A. (1981). *Materialism and Society in the Mid-Eighteenth Century: La Mettrie's Discours Préliminaire*. Droz, Geneva/Paris.

Thomson, A. (2008). *Bodies of Thought. Science, Religion, and the Soul in the Early Enlightenment*. Oxford University Press, Oxford.

Torero-Ibad, A. (2009). *Libertinage, science et philosophie dans le matérialisme de Cyrano de Bergerac*. Honoré Champion, Paris.

Vandewalle, B. (2011). *Spinoza et la médecine. Éthique et thérapeutique*. L'Harmattan, Paris.

Vernière, P. (1954). *Spinoza et la pensée française avant la Révolution*. PUF, Paris.

Westphal, J. (2016). *The Mind-Body Problem*. MIT Press, Cambridge.

Wolfe, C. (2010). Endowed molecules and emergent organization: The Maupertuis-Diderot debate. *Early Science and Medicine*, 15, 38–65.

Zac, S. (1963). *L'idée de vie dans la philosophie de Spinoza*. PUF, Paris.

8

The Emergence of Transformism in the 18th and 19th Centuries

Stéphane TIRARD
CFV, Nantes Université, Nantes, France

8.1. Introduction

> Who can doubt that our birds, which soar through the air, came from the volatile genus of fish and that our land animals, which have neither the disposition to fly nor the ability to rise above the earth, came from those that crawl on the sea floor? (Maillet 1984, p. 249)

This statement, taken from a philosophical tale written in the early 18th century by Benoît de Maillet (1656–1738) and published in 1748 under the title *Telliamed, or Conversations of an Indian Philosopher with a French Missionary*, reflects ideas that could be described as transformist. Their divergence from traditional Christian thought explains the posthumous publication and the choice of title, *Telliamed*, which is an inversion of the author's name, as a pretense to maintain relative anonymity. In this long dialog between an Indian and a missionary, Maillet uses the former to present an account of the history of creation that does not correspond to that of Genesis. Maillet imagines that life came from space in the form of germs after the earth had cooled sufficiently and that the vapor condensed to form seas. These germs of life present in the oceans were able to grow and give rise to marine plant and animal species. After the sea receded, land species emerged from the sea; flying fish, for example, transformed into birds. The world conceived by Maillet is

History of Life Sciences and Medicine 2,
coordinated by Céline CHERICI. © ISTE Ltd 2026.

therefore evolving, and he devotes his book to describing it. However, this approach was exceptional in the 18th century, as it was dominated mostly by a fixed way of thinking.

Most naturalists recorded their work with a fixist approach. In a century focused on classification, the major work of the Swedish scientist Carl Linnaeus (1707–1778) illustrates this desire to study the diversity of the world. The naturalist from Uppsala devoted his life to this task, constantly expanding the scope of his work to include all plants and animals, as well as minerals. His classification, known as artificial because it favored specific criteria, aimed to describe the order of species created by God at the beginning of the world and subsequently remained unchanged, although he considered that variations could appear within them. This conception of the world, which can be described as creationist because it respects the dogma of Genesis and fixist because species are immutable, remained widely shared during the Enlightenment.

The aim of this chapter is to examine the theories that emerged during this period, which, like Maillet, considered other possibilities that diverged from the constraints imposed by biblical Genesis. These are scientific theories, known as transformist theories, which, in the 18th century, claimed that living species had not remained unchanged since the primordial creation, as conceived by Christian religious dogma. The term "evolutionism" could also have been chosen, as it can encompass the theories that interest us here. However, at the time of this study, the term "evolution" referred to the phenomena of embryo and fetal development, and its use in the context of species transformation was applied to later theories, notably that of Charles Darwin.

As France was the main setting for the emergence of these ideas, we will explore the 18th and early 19th centuries, focusing on the works of authors such as Benoît de Maillet, Bonnet and Diderot. However, we begin with Buffon, whose work, although not transformist, contains concepts that reveal his profound reflection on the limits of fixity. Our discussion will then revolve around the first quarter of the 19th century, when transformist theories emerged, such as those of Lamarck and Geoffroy Saint-Hilaire, and debates arose with Cuvier, who remained attached to fixity.

8.2. Buffon and listing of species over time

First, it is important to note that Buffon was not a transformist. Although he repeatedly questioned the diversity of living beings throughout his work, he was careful to first theorize about species and then about families as fixed entities, with no possible transition from one to the other. The French naturalist's work is nonetheless imbued with an important conceptual ambition that deserves to be analyzed; although it is this ambition that gives his work its originality, it is also characteristic of a desire to think about the history of the Earth and living beings outside the confines of imposed dogmas[1].

8.2.1. *A work of concepts*

A member of the Burgundian bourgeoisie, Georges-Louis Leclerc, Comte de Buffon (1707–1788), joined the Royal Academy of Sciences in 1733 as an assistant in the mechanics section, owing to his work in mathematics. In 1744, he joined the botany section after becoming intendant of the King's Garden and Natural History Cabinet on July 26, 1739. This important position helped make him a prominent figure. However, his reputation was established mainly by his published work, which was divided into several volumes comprising his *Histoire naturelle* (*Natural History*):

– *Natural History, General and Particular*, 15 volumes, 1749–1767;

– *Natural History of Birds*, 9 volumes, 1770–1783;

– *Natural History, General and Particular, Supplement*, 7 volumes, 1774–1789;

– *Natural History of Minerals*, 5 volumes, 1783–1788.

His work is characterized by three main aspects. First, it is a descriptive work of natural history based on an extensive compilation and supplemented by his own observations. Buffon produced this vast work with the help of collaborators, notably Louis Jean-Marie D'Aubenton (1716–1799), who wrote important parts of this collection of works. It is also a literary work, as Buffon had real talent for writing. These first two points undoubtedly contributed to the remarkable posterity of Buffon's work, and there were countless reissues of his *Natural History* in the 19th century.

Lastly, and this is an important point to emphasize, it is crucial work in terms of the concepts it contains. It is essential to view them in their entirety, because it is by linking them together that their coherence in Buffon's work becomes apparent.

1 For more information on Buffon, we refer to the works of Roger (1988, 1989) and Schmitt (2023).

The first concept was presented in the second volume of *Natural History*, published in 1749. These are the organic molecules make up all living beings. He first presents them as living, primitive and incorruptible organic parts (Buffon 1748, Vol. 2, p. 24) and quickly refers to them by their definitive name, *organic molecules*. According to him, these entities are the building blocks of both animals and plants. They are returned to the soil after the decomposition of all kinds of corpses, where they are reabsorbed by plants and subject to further transfer to animals and then from animals to animals through food. This is a true cycle that is therefore continuously at work in nature. Organic molecules also play a role in reproduction, as discussed below. At the end of this presentation of organic molecules, Buffon presents a central principle that will ultimately prove crucial in his history of life on Earth in regard to considering the formation of organic molecules themselves: "The most ordinary work of Nature is the production of organic matter, which is its most familiar action" (Buffon 1748, Vol. 2, p. 38).

The *internal mold*, another Buffonian concept, is also crucial (Buffon 1748, Vol. 2, p. 34 *et seq.*). The naturalist asserts that each species is characterized by a particular organization, preserved from generation to generation by organic molecules that take the imprint of this mold and are transmitted through the seed of both parents[2]. The offspring is thus produced in the image of its parents. Adding to this concept is that of the *external model*, which is, in a way, complementary to it. This is the characteristic form of the organism in question. The internal mold and the external model characterize the *prototype* of the species. Thus, the prototype of the horse, given as an example in Volume IV, was the first horse from which all horse breeds descended. Buffon noted, however, that there are variations within the limits of the species' characteristics over generations:

> The first animal, the first horse, for example, was the external model and internal mold on which all horses that were born, all those that exist, and all those that will be born, were formed; but this model, of which we know only copies, has been able to alter or perfect itself by communicating its form and multiplying. The original imprint remains intact in each individual; but although there are millions of them, none of these individuals is, however, entirely similar to another individual, nor, consequently, to the model of which it bears the imprint: this difference proves how far Nature is from doing anything absolute, and how well it knows how to nuance its works. (Buffon 1753, Vol. IV, p. 216)

2 From sexual maturity onward, this seed is the result of excess organic molecules accumulated through nutrition.

Buffon's definition of *species* either explicitly or implicitly incorporates the preceding concepts and deserves to be analyzed in detail.

8.2.2. *Defining the species and its boundaries*

Buffon gives his definition of species in a famous passage from the article devoted to donkeys:

> An individual is a separate, isolated being, detached from others and having nothing in common with them, except that it resembles them or differs from them: all similar individuals existing on the surface of the earth are regarded as constituting the species of these individuals; however, it is neither the number nor the collection of individuals that constitutes the species, but rather the constant succession and uninterrupted renewal of these individuals; for a being that would last forever would not be a species, nor would a million similar beings that would also last forever: the species is therefore an abstract and general word, the thing itself existing only when considering Nature in the progression of time and in the constant destruction and equally constant renewal of beings. (Buffon 1753, Vol. IV, pp. 384–385)

By insisting that species must be considered over time and through their perpetuation, Buffon introduces time into his definition and links this concept to the question of reproduction; in other words, the transmission of species traits. He noted that the ability to produce fertile offspring together is the prerogative of individuals of the same species; conversely, if individuals resulting from crossbreeding are not interfertile, their parents do not belong to the same species. Thus, donkeys and horses are not of the same species, "since they only produce defective and infertile individuals when they reproduce with one another" (Buffon 1753, Vol. IV, p. 384). This boundary between species is important to Buffon, and he discusses it at length in this article. While he understood that a certain variety could exist within a species, he nevertheless believed that one species could not transform into another. Transformations that were possibly reversible, in the sense of degeneration, did occur within species, particularly as a result of domestication, but this diversification remained limited to a fixist framework. It should be emphasized how consistent this conception of the species is with the concepts presented above. Indeed, it is around the species prototype that variations occur, and thus, ancestral traits or a few variations acquired over generations characterize the external model and the internal mold and are transmitted through the imprints carried by organic molecules.

Buffon later extended the concept of interfertility to the family[3] and considered that there was a form of kinship between the species it contained, with the latter constituting a diversification resulting from degeneration. However, the naturalist remained a fixist, since one family cannot give rise to another family.

8.2.3. *Linking the history of the Earth and life:* Epochs of Nature

In 1778, 30 years after the publication of the first volume of his *Natural History*, entitled *Theory of the Earth*, Buffon published Volume V of the supplements, which contained *Des Époques de la Nature* (*The Epochs of Nature*), a profound revision of his conception of the history of the Earth and living beings. One of the most important aspects of this vast narrative, divided into seven epochs, is undoubtedly the irreversibility of the process described, which is due to the gradual and irremediable cooling of the Earth since its formation. Thirty years later, Buffon moved from a cyclical history of the Earth to a sagittal history, with the progression of time resulting from the irreversible transformation of conditions (Gohau 1990, pp. 106–108). *The Epochs of Nature* are therefore not simply an extension of the *Theory of the Earth* described in the first volume of *Natural History*, but rather a new theory that gives time a determining function.

According to Buffon, the entire process took approximately 75,000 years. Without dwelling here on the naturalist's calculations and the debates surrounding this timeframe, Buffon estimated the age of the Earth by modeling the cooling of the planet using iron balls heated to very high temperatures in the forges of his Burgundian estate of Montbard and measuring the time it took for them to return to room temperature, then extrapolating the phenomenon to a body the size of the Earth (Buffon 1988, pp. LX–LXVII). The naturalist's history of the Earth can be summarized as follows. During the *First Epoch*, the Earth was a flattened, incandescent sphere that solidified during the *Second Epoch*. Buffon therefore drew on his work to estimate the age of the Earth. The roughness and bubbles observed on the balls were associated with mountains and caves, respectively. During the *Third Epoch*, water vapor in the atmosphere condensed into a primitive ocean. With only the mountain peaks remaining above the water, it was during this stage that life appeared. The *Fourth Epoch* was marked by the seepage of water into the caves. When the roofs of these caves collapsed, seas formed. The large quantities of vegetation that had developed on the continents were swept away by the waters, settling at the bottom of the seas and forming coal, which ignited and gave rise to the first volcanoes. During the *Fifth Epoch*, animals formed spontaneously on land

3 Buffon counted 38 families of quadrupeds.

that was particularly fertile at the time, as it was rich in organic molecules. The northern regions, which received less sunlight, were the first regions to reach the ideal temperature for this genesis during the cooling period. The *Sixth Epoch* marked by the separation of continents, as indicated by the presence of identical animals on either side of the oceans formed during the collapses. Finally, the *Seventh and last Epoch* corresponds to the period when "the power of man assisted that of nature" (Buffon 1988, p. 206, 235[4]).

8.2.4. *Spontaneous generations at the beginning of species*

Jacques Roger showed that between 1749 and 1778, when *Epochs of Nature* was published, Buffon moved from a Newtonian physical conception to a Cartesian and chemical model, which, according to him, allowed naturalists to consider the origin of organic matter (Buffon 1988, p. LXVIII). As for the question of the origins of life, he rightly noted that the origin of organic molecules is difficult to identify in *Epochs of Nature*. Buffon thus protected himself from criticism by theologians, and it is therefore necessary to gather dispersed evidence to reconstruct the process of the emergence of life as conceived by Buffon. Before this collection, it is worth returning to Buffon's organic molecules, which play a crucial role in the conception of life. The history of life presented in *Epochs*, and contextualized in this history of the Earth, calls for a foundation. Regarding the origin of these organic molecules, Buffon asserts that they were formed by the action of heat on the substances that compose living beings, that is, "aqueous, oily, and ductile matter" (Buffon 1998, p. LXVIII). In the *Fifth Epoch*, the process is described in some insightful detail:

> [...] for all the aqueous, oily, and ductile parts that were to enter into the composition of organized beings fell with the waters onto the northern parts of the globe much earlier and in much greater quantities than onto the southern parts: it is in these aqueous and ductile materials that living organic molecules began to exercise their power to shape and develop organized bodies: and since organic molecules are only produced by heat on ductile materials, they were also more abundant in the northern lands than they could be in the southern lands, where these same materials were in smaller quantities. It is not surprising that the first, strongest, and greatest productions of living nature took place in these same northern lands; whereas in those of the equator,

4 The quotations were taken from the critical edition of *The Epochs of Nature* by Jacques Roger (Buffon 1988). We therefore give the page numbers in the critical edition and then in the work itself.

> and particularly in those of South America, where the quantity of these same ductile materials was much less, only inferior species were formed, smaller and weaker than those of the northern lands. (Buffon 1988, pp. 157–158, 185–186)

It was also during this *Fifth Epoch*, in which Buffon specified that 35,000 or 36,000 years ago, the Earth was supposed to have cooled enough to support life:

> Everything that exists today in living nature could have existed in the same way as soon as the temperature of the Earth became the same. (Buffon 1988, p. 138, 165)

The rest is clear if we return to the *Third Epoch*:

> Here, we must represent to ourselves the course of nature, and even recall the idea of its means. Living organic molecules existed as soon as the elements of gentle heat were able to combine with the substances that make up organized bodies; they produced an infinite number of plants on the higher parts of the globe, and an immense number of shellfish, crustaceans, and fish in the waters, which soon multiplied by means of generation. This multiplication of plants and shellfish, however rapid we may suppose it to have been, could only have taken place over a great number of centuries, since it produced volumes as enormous as those of their remains; for, in order to judge what has happened, we must consider what is happening. It takes many years for oysters, which accumulate in certain places in the sea, to multiply in sufficient quantities to form a kind of rock. And how many centuries did it take for all the calcareous matter on the surface of the globe to be produced? Are we not forced to acknowledge that it not only took centuries, but centuries upon centuries for these marine products to not only be reduced to powder, but transported and deposited by the waters in such a way as to form chalk, marl, marble and limestone? (Buffon 1988, pp. 101–110, 113)

This is how ammonites date back to the earliest days of water on Earth, followed by elephants and rhinoceroses, in other words, all large animals. Their large size can be explained by the power of nature that reigned over the Earth at that time:

> Nature was then in its prime and was working on organic and living matter with greater power in a warmer climate: this organic matter was more divided, less combined with other materials, and could

come together and combine with itself in larger masses to develop into larger dimensions: this cause is sufficient to explain all the gigantic productions that appear to have been frequent in these early ages of the world. (Buffon 1988, pp. 77–78, 99)

Large species formed first in northern regions, as these areas were the first to reach a compatible temperature, with their formation occurring as the globe cooled. This reminds us that, according to the naturalist, all major stages in the development of life on Earth are dependent on the irreversible course of this decrease in temperature.

Buffon's work is therefore marked by the evolution of his own thinking, and his final conception of the Earth reveals the importance he attached to the introduction of the sagittal time scale, on which successive irreversible geological states have been recorded. On Earth, with its irrevocably changing climate, species appeared as a result of the fertility of matter and the abundance of organic molecules. Families of living beings are nevertheless subject to variation but remain independent of one another. Therefore, Buffon's conception of nature involves an evolving Earth that produces a fixed living world, albeit one that is susceptible to variation through degeneration.

8.3. Thinking about life beyond dogma: a plethora of approaches

While fixist thinking is widely shared, other naturalists besides Buffon and philosophers put forward proposals introducing new approaches to the history of the Earth and species. For them, it was a matter of allowing themselves to split from the biblical dogma of Creation and to open up a reflection on other possibilities regarding the beginning of life on Earth, possibly in relation to its history, and on the possibility of the subsequent transformation of species.

8.3.1. *The chain of beings: expression of the principle of plenitude*

Following Leibniz and Charles Bonnet, many 18th-century naturalists adopted the ancient principle of plenitude, according to which the world was as complete as possible, with no empty spaces between species. In his *Traité d'insectologie* (Bonnet 1745), this is how Bonnet represented a chain of beings that would mark the naturalist thinking of the time.

It is a gradual progression from humans to "more subtle matter", passing through a series of animals, then plants, followed by minerals, and ending with the four elements: earth, water, air and fire. For the author, this order, from the most complex at the top to the simplest at the bottom, does not imply the possibility of transformation from one level to the next, as the framework conceived by Bonnet remains firmly fixed.

Although it is widely accepted, this view of the world order nevertheless encountered opposition, notably from Voltaire, who, in the article "*Chaîne des êtres créés*" (Chain of Created Beings) in his *Dictionnaire philosophique*, denied this "alleged gradation" that he took from Plato, contesting what some imagined as the "imperceptible passage from raw matter to organized matter, from plants to zoophytes, from these zoophytes to animals, and from animals to man", which could lead one to believe that by passing through geniuses, it would be possible to thus rise to God. However, the philosopher asserts that there is infinity between God and "the most perfect of geniuses" created by him and states the following: "Is there not an obvious gap between the monkey and man?" (Voltaire 1964, pp. 107–108).

8.3.2. *Maupertuis: errors producing new species*

In the mid-18th century, mathematician Pierre Louis Moreau de Maupertuis (1698–1759) formulated his views on the formation of living bodies. In 1746, in *Vénus physique*, he explained how, in his opinion, organization is transmitted from one generation to the next:

> If this force exists in nature, why should it not occur in the formation of the bodies of animals? In each seed, let there be parts destined to form the heart, the head, the entrails, the arms, the legs; and let these parts each have a greater connection with the part which, for the formation of the animal, must be its neighbor, than with any other; the heart will form: and even if it were a thousand times more organized than it is, it would still form. (Maupertuis 1997, p. 78)

He also raises questions about the origin of species diversity, founding his argument on the idea of an indivisible part of the body that is the essence of the animal, and concludes by stating:

> Could this part only ever reproduce an animal of the same species? Or could it produce all possible species, solely through the diversity of combinations of the parts with which it would combine? (Maupertuis 1997, p. 110)

In 1754, in his *Essai sur la formation des corps organisés* (*Essay on the Formation of Organized Bodies*), he proposed an explanation on the basis of the idea that God, in creating the world, endowed every part of matter with perception. Thus, during generation, the transmitted matter, remembering its former disposition, would know how to order itself to transmit the same organization as that of its parents to its offspring, something that a simple attraction between parts of matter could not produce. Maupertuis thus explained the resemblance within species as well as the monstrosities or the possibility of the formation of *mixed races* that were sterile. However, by the same principle, he also explains how new species can form:

> Could this not explain how, from just two individuals, the multiplication of the most dissimilar species could have ensued? Their initial origin would have been due to a few fortuitous productions, all of which would have failed to retain the order of the elementary parts found in their parent animals: each degree of error would have produced a new species; and repeated deviations would have led to the infinite diversity of animals that we see today, which may increase further with time, but which would only bring imperceptible increases over the course of centuries.

Maupertuis also dates back to the time when all matter on Earth was in the same fluid state and suggests the following:

> The least active parts of matter will have formed metals and marble; the most active parts will have formed animals and humans.

Hardened matter cannot, therefore, form new products, unlike living beings, whose matter retains its fluidity. On the basis of these principles, Maupertuis formulated an original proposition regarding the transformation of matter as well as species.

8.3.3. *Diderot and the animalization of matter*

Denis Diderot (1713–1784) developed his thinking on natural history and medicine through countless books, and he formalized the results of his reflection in influential texts that reveal the originality of his thought: his articles in the *Encyclopédie*[5], his *Letter on the Blind* (1749), his conversations, including

5 In his article "Animal" in *Encyclopédie*, the philosopher explicitly reflects Buffon's theories.

the famous *d'Alembert's Dream* (1769), and, lastly, *Elements of Physiology* (1778–1780), an original synthesis revealing a conception of the transformation of species.

The philosopher adheres to the idea of spontaneous generations of microscopic beings, as these are merely an expression of the transformation of matter. He also asserts that "the transition from a state of inertia to a state of sensitivity, and spontaneous generations" are two phenomena that must be understood. In his *Elements of Physiology*, he offers reflections that extend those he had already mentioned in his *Entretiens* (Discourses) and generally confirms how, in his thinking, the problems of the origin of living matter, its transformation, epigenesis, spontaneous generations and the origin of living beings are inseparable. All these concepts are linked by an idea of transformation:

> Nature has only created a very small number of beings, which it has varied infinitely, perhaps only one by combination, mixture, or dissolution, from which all others have been formed. (Diderot 1994, p. 1261)

Epigenesis is discussed in great detail, both in *Entretien entre d'Alembert et Diderot* (*Discourse between d'Alembert and Diderot*), in which the former, taken as an example, is the subject of sarcasm, and in the *Elements of Physiology*, when Diderot describes the origin of the germ:

> The scattered molecules that must form the germ necessarily gather there; once gathered, they form a seed; this seed has only one development; it is a tree, and so it is with animals and humans. (Diderot 1994, p. 1274)

Epigenesis and spontaneous generations therefore fall under the same necessary principle of material transformation. As we will see later, Lamarck also compared them and used the former, by analogy, to prove the existence of the latter. Notably, the philosopher ultimately accepted an absolute beginning to the history of life:

> Vegetation, life, or sentience, and animalization are three successive operations. The plant kingdom could well be and have been the primary source of the animal kingdom, which in turn originated in the mineral kingdom, which itself emanated from heterogeneous universal matter. (Diderot 1994, p. 1262)

Diderot therefore considers both the origin of living matter and the successive transformation of species.

8.4. Lamarck, story of a conversion: from fixism to "transformism"

The work of Jean-Baptiste Lamarck (1744–1829) is marked by two distinct periods. The early years of the 19th century were a time of conversion for him: he abandoned fixism and developed and published an explicitly transformist theory in 1802.

8.4.1. *Meteorology, hydrogeology and biology*

After beginning his career in the military, Lamarck turned to botany and the life sciences. With Buffon's support, he published his *Flore Françoise* in 1779. Lamarck's classification method differed from that of the Swedish naturalist Linnaeus, which greatly pleased Buffon. However, the naturalist was not content with the fame that his flora brought him and turned his attention to chemistry, a field in which he struggled to establish ideas that would never consider the results of Lavoisier's chemical revolution. It was not until 1794 that he published his *Recherches sur les causes des principaux faits physiques* (*Research on the Causes of the Principal Physical Phenomena*). His interest in chemistry was far from anecdotal because, according to his chemical concepts, living beings produced all the combinations that made the matter around us possible. This idea obviously ruled out any conception of the beginning of life, since, according to this principle, no combined matter could pre-exist life. For the same reasons, spontaneous generations could not be accepted by Lamarck, since matter cannot spontaneously combine and become more complex.

In 1793, Lamarck was appointed a professor of zoology of invertebrate animals at the French National Museum of Natural History. This marked a shift in his focus, which resulted in the publication of a comprehensive work in 1801 entitled *Système des animaux sans vertèbres* (*System of Invertebrate Animals*). The following year, in 1802, he published two works, *Hydrogéologie ou recherches sur l'influence qu'ont les eaux sur la surface du globe terrestre* (*Hydrogeology, or Research on the Influence of Water on the Earth's Surface*), a general approach to the transformations of the Earth's surface under the influence of water, and *Recherches sur l'organisation des corps vivants* (*Research on the Organization of Living Bodies*). Although these two themes seem unrelated, it should be noted that, for Lamarck, they are linked by the following:

> Good terrestrial physics must include all considerations of the first order relating to the Earth's atmosphere; then all those of the same kind concerning the state of the outer crust of this globe, as well as the

> modifications and changes it continuously undergoes; lastly, those of the same kind belonging to the origin and development of the organization of living bodies. Thus, all these considerations naturally divide terrestrial physics into three essential parts, the first of which must include the theory of the atmosphere, meteorology; the second, that of the outer crust of the globe, hydrogeology; and the third, that of living bodies, biology. (Lamarck 1802b, p. 8)

Although Lamarck's naturalist works at the end of the 18th century were oriented toward invertebrates, in line with the subject of his career, he nevertheless aspired to a broad understanding of nature, both inanimate and living, and linked these two states. It is important to emphasize the consequences of the chemical concepts that Lamarck developed in his early work. Thus, in *Mémoires de physique et d'histoire naturelle*, he did not hesitate to assert that living bodies are the primary cause of the existence of all the compounds (Lamarck 1797, p. 238 ff.). He added that "all the compounds observed on our globe are due, either directly or indirectly, to the organic capacities of living beings" (Lamarck 1797, p. 386). As indicated above, under these conditions, the primordial beginning of living bodies, and therefore spontaneous generations, is impossible.

8.4.2. *Spontaneous generations, a model of the transformation process*[6]

It is also worth taking a look at his *System of Invertebrate Animals* (Lamarck 1801) to discover some of his zoological concepts. Lamarck describes the classes of invertebrate animals: mollusks, crustaceans, arachnids, insects, radiates and polyps. The latter, he tells us, are animals with no head and no internal organs other than an intestinal canal, the entrance of which serves as both the mouth and anus. Among them, we distinguished serrated polyps, rotifer polyps and amorphous polyps. These are the simplest polyps, lacking organs, and include vibrios, proteas and monads. At the end of this series, he describes the simplest species, *Monas termo*, and specifies the following:

> Here we have the smallest and simplest of all animals, the apparent end of animality, which naturalists have finally managed to discover in the animal kingdom and classify methodically. (Lamarck 1801, p. 397)

6 Regarding this subject, see Tirard (2006).

However, a paradox must be highlighted. In his *Research*, Lamarck always supports the same chemical theory. He refers to his *Mémoires de physique d'histoire naturelle* (*Memoirs of Natural History Physics*). The appendix on "species among living bodies", for example, contains explicit statements showing that he still believes that the waste of living bodies undergoes changes that provide "the materials of all mineral substances". His chemistry therefore remains present in the theory he proposes. This gives rise to paradoxes that Lamarck does not seem to notice. Indeed, according to the principles of his chemistry, gelatinous bodies cannot be the result of an organization of mineral matter without the intervention of living beings.

In 1802, on the 27th day of Floréal in year X (May 16, 1802), Lamarck gave his annual opening lecture at the museum, which was published the same year in a work entitled: *Research on the organization of living bodies, particularly on their origin, the cause of their development and the progress of their composition, and on that which, continuously leading to destroy them in each individual, necessarily leads to their death; Preceded by the opening speech of the course on zoology, given at the French National Museum of Natural History, in year X of the Republic.*

With this text, the naturalist fully entered into his new theory. Let us follow the course of his argument to grasp the rhetoric of his reasoning. The text opens with the following title: "Progress in the composition of the organization of living bodies, as circumstances favor them". The notion of progress and the idea of change are thus clearly stated. A subsequent title mentions that there is a "degradation of organization from one end of the animal chain to the other". Following this "degradation", Lamarck indicates that he is "taking the order in the opposite direction to that of nature" (Lamarck 1802, p. 14). In other words, nature did not follow the traditional order of the chain of beings from the most complex to the simplest but rather progressed from the simplest to the most complicated.

Lamarck therefore follows a series of "destructions". After mammals, birds, reptiles and fish, he observes the disappearance of the spine, which leads to mollusks, annelids and crustaceans. Then, the heart disappears, which arachnids and insects do not have. Fertilization then disappears, leading to worms, and the disappearance of sight leads to radiata. What remains are polyps, which only retain touch and a food canal. At the end of this chain, the last order of polyps, the animalcules "are nothing more than animalized points, gelatinous corpuscles, transparent, of a very simple shape, and contractile in all directions" (Lamarck 1802, p. 36). These are the simplest polyps, as he described them in his *System of Invertebrate Animals*, and this is the moment he chose to assert a new point that was considerably different from his previous theses:

> It is undoubtedly among them (polyps) that we find the first rudiments of animal life directly produced by nature, in a word, spontaneous generations. (Lamarck 1802, pp. 36–37)

Thus, spontaneous generations could exist, despite the constraints of the chemical principles conceived by the naturalist. There appears to be a major shift in the evolution of Lamarck's ideas.

Spontaneous generations are therefore given a very significant role in the naturalist's argument because, by describing their mechanism, he actually describes the first stages of species transformation and reveals the fundamental principles of his theory. Indeed, in his description of primordial mechanisms, Lamarck introduces the factors necessary for his explanation of species transformation. During the animalization of matter, in other words, during the formation of the first animalcule, gelatinous matter, that is to say, matter in "the state closest to the fluid state, consequently having the greatest flexibility and only sufficient consistency to be capable of forming the containing parts" (Lamarck 1802, p. 106), is first subjected to uncontrollable fluids (heat and electricity). It then enters a state of "vital orgasm":

> A particular tension in all points of the soft parts of these living bodies, which holds their molecules at a certain distance from one another […], which they are liable to lose, by the simple effect of attraction, when the cause that maintains it ceases to act. (Lamarck 1802, p. 79)

Next, containable fluids (liquids and gases) act on matter in this state and mold it "through their movements and other influences, developing it and, with time and all favorable circumstances, complicating and perfecting it" (Lamarck 1802, pp. 107–108). This is where one of the most fundamental concepts of Lamarckian theory comes into play: *habit*. The fluids dig into the gelatinous mass of matter, forming a dimple:

> Imperceptibly, this dimple, through habit of filling up and through more frequent use of its pores, will gradually increase in depth; it will soon take the form of a pocket, or a tubular depression, porous in its walls, or blind, with just one outlet. This is the first alimentary canal created by nature, the simplest organ of digestion. […] This is indeed the structure of the food canal of all animals of the polyp class. (Lamarck 1802, p. 112)

It is therefore possible to consider that the description of this phenomenon, on the scale of spontaneous generations, in a way models the mechanisms that Lamarck later implemented on the scale of more complex species[7]. The action of the environment, habit, a very long timeframe, and the action of fluids will always be fundamental to his explanation. One difference in these more highly evolved animals is that the fluids are inside the organism and gradually transform through their internal action. This is illustrated, for example, in this passage on the production of horns from *Philosophie zoologique* (*Zoological Philosophy*), published in 1809:

> In their frequent fits of anger, especially among males, their inner feelings, through their efforts, direct the fluids more strongly towards this part of their head, and there is a secretion of horny matter in some, and of bony matter mixed with horny matter in others, which gives rise to solid protrusions: hence the origin of the horns and antlers with which most of these animals are armed. With regard to habits, it is curious to observe the result in the particular shape and size of the giraffe (camelopardalis): it is known that this animal, the largest of the mammals, inhabits the interior of Africa, and that it lives in places where the earth, almost always arid and without grass, forces it to graze on the foliage of trees, and to continuously strive to reach it. As a result of this habit, which has been sustained for a long time in all individuals of its species, its front legs have become longer than its hind legs, and its neck has become so elongated that the giraffe, without standing on its hind legs, raises its head and reaches a height of six meters (nearly twenty feet). (Lamarck 1809, pp. 256–257)

8.4.3. *The successive formation of species*

In *Zoological Philosophy*, he formulated two laws, which he expanded into four laws in the *Natural History of Invertebrate Animals* (Lamarck 1815–1822). These laws cover changes in organs according to the principles of use and disuse, as well as their transmission to subsequent generations:

---

7 It should also be noted that Lamarck, considering spontaneous generations to be too subtle a phenomenon to be directly observable in nature, asserts that they are analogous to fertilization: "By comparing analogous facts, it would be very possible for me to show that what happens in sexual fertilization is imitated by nature in certain circumstances, and is itself brought about by nature on isolated masses of matter that are in a state suitable for undergoing this effect" (Lamarck 1802, p. 101).

> First law
>
> In any animal that has not exceeded the limit of its development, the more frequent and sustained use of any organ gradually strengthens that organ, develops it, enlarges it, and gives it a power proportional to the duration of that use; while the constant lack of use of such an organ weakens it imperceptibly, deteriorates it, progressively diminishes its capacities, and eventually causes it to disappear.
>
> Second law
>
> Everything that nature has caused individuals to acquire or lose through the influence of circumstances to which their race has long been exposed, and consequently through the influence of the predominant use of a particular organ, or through the constant lack of use of a particular part; it preserves it by transmission to the offspring of those individuals, provided that the acquired changes are common to both sexes, or to those that produced these new individuals. (Lamarck 1802, 1809, p. 235)

These laws explain how transformations that have become fixed over generations have led to the formation of lineages. One class leads to the next, sometimes giving rise to dichotomies that break with the strict linearity of the formulation from 1802. Furthermore, in 1809, Lamarck anchored his lineages in two centers of spontaneous generation. One began with worms and led to vertebrates, and the other began with infusoria and was limited to polyps and radiata.

Lamarck's name is often associated with the formulation of the principle of inheritance of acquired traits. It should be noted that he never referred to it as such himself and this phrase only came into use later in the 19th century (Gayon 2006). Although this principle can indeed be found in Darwin's work, it is not sufficient to establish any connection between the two theories. Darwin was not inspired by Lamarck in any way. Lamarck based his theory on the plasticity of living matter under the influence of circumstances and habits, whereas the English naturalist based his theory on variation and selection.

During the first two decades of the 19th century, Lamarck's theories spread within the rich naturalist context of French science. Lamarck's colleagues at the Museum included Georges Cuvier (1769–1831) and Étienne Geoffroy Saint-Hilaire. What would later be called "Lamarckian transformism" found a relatively limited audience and a few followers who, like Jean-Baptiste Bory de Saint-Vincent

(1778–1848), promoted it. After Lamarck's death, the transformist movement continued, albeit in competition with Cuvier's ideas.

8.5. Cuvier and the compositional plans

Cuvier was born in 1769 in Montbéliard, then part of the Duchy of Württemberg, and studied at the academic college in Stuttgart. In 1788, he became a tutor in Normandy for six years. In 1794, he was noticed by Henri-Alexandre Tessier (1741–1837), a physician and naturalist, who recommended him to Étienne Geoffroy Saint-Hilaire, whom he quickly joined in Paris. He assisted Daubenton and Lacepède and, in 1795, became deputy professor of comparative anatomy at the Museum. There he reorganized the anatomy cabinet and pursued a prestigious career at the Museum and the Collège de France (1799), helping to found both comparative anatomy and vertebrate paleontology[8]. He formulated a set of ideas that would shape the life sciences for several decades. The "Preliminary Discourse" he published in 1812 in his *Recherches sur les ossements fossiles* (*Researches on Fossil Bones*) explains two of the main principles that structured all his research. His extensive work in comparative anatomy, both on animals and fossils, some of which he showed to be the remains of extinct species, led him to propose the principle of organ correlation:

> Fortunately, comparative anatomy possessed a principle which, when well developed, was capable of clearing all difficulties: that of the correlation of forms in organized beings, by means of which each kind of being could, in theory, be recognized by each fragment of each of its parts.
>
> Every organized being forms a whole, a unique and closed system, in which all parts correspond to each other and contribute to the same definitive action through mutual reaction. None of these parts can change without the others also changing; and consequently, each of them, taken separately, shows and provides all the others. (Cuvier 1812, p. 58)

This principle of correlation, which he aspires to elevate to the status of a scientific law, enables him to reconstruct species from a few anatomical elements, or even, in principle, from a single element. In doing so, he demonstrates the past existence of species that are now extinct and explains these changes using his

8 Cuvier also pursued a political career (1813, master of requests; 1814, councillor of state; 1832, peer of France) and wielded considerable institutional power.

second principle, which involves a series of successive catastrophes that, according to him, decreased in intensity over the course of these successive events, with the most recent corresponding to the flood. These catastrophes, which he finds evidence of in sedimentary layers, are what he believes caused the disappearance of species, which were then replaced by others from elsewhere on the globe. These are therefore not new creations but rather disappearances followed by new distributions.

> Life has therefore often been troubled on this earth by terrible events; calamities which, in the beginning, may have shaken the entire envelope of the planet, but which have since become less profound and less widespread. Countless living beings have been victims of these catastrophes; some were destroyed by floods, others were left high and dry when the sea floor suddenly rose; their very races have died out forever, leaving just a few fragments behind, barely recognizable to the naturalist. (Cuvier 1812, pp. 11–12)

Cuvier therefore does not consider the transformation of one species into another; although he believes that species can become extinct, he does not believe that they change over time. He therefore remains faithful to the fixist framework. The place left vacant by a missing species will subsequently be occupied by another species that already exists elsewhere. If species have a history, it is one without transformation, and the only event that marks them is their eventual extinction, which is recorded in the history of the Earth by the event that causes it.

8.6. Geoffroy Saint-Hilaire and the principle of recapitulation[9]

Historiography has given too little attention to the work of Étienne Geoffroy Saint-Hilaire. As a contemporary of Lamarck, he was, like him, appointed professor at the French National Museum of Natural History when it was founded, where he was given the position of professor of zoology (quadrupeds, cetaceans, birds, reptiles and fish), which was quickly divided into two. In 1794, he became a professor of zoology (mammals and birds), whereas the other position of professor of zoology (reptiles and fish) was given to Étienne de Lacepède in 1795. After returning from the Egyptian campaign, Geoffroy Saint-Hilaire continued to establish himself as a comparative anatomist and became involved in experimental embryology, teratology and paleontology. His scientific work was highlighted by a fundamental thesis that would influence all of his work: the unity of composition.

9 We refer readers to the texts by Geoffroy Saint-Hilaire, reproduced in Hervé Le Guyader's book (Le Guyader 1998) and to his comments, on which we have based our own remarks.

This concept, which he gradually extended to the entire animal kingdom, led him to formulate the principle of recapitulation, asserting that "a 'higher' animal passes through stages during its development that correspond to the adult states of so-called 'lower' animals" (Le Guyader 1998, p. 17). Geoffroy thus linked species in a long process of successive transformations, with the most advanced retaining traces of this in their development.

Geoffroy's thesis, based on a single compositional plan, was in fact opposed to that of Cuvier, who proposed four. From the 1820s onward, deep opposition developed between the two former friends, and successive events served only to exacerbate their rivalry, such as when Geoffroy created a new intermediate genus between reptiles and mammals to classify a gavial fossil found near Caen, thus sparking the "analogies controversy". However, the controversy broke out publicly when Laurencet and Meyranx established an analogy between cephalopods and vertebrates, which Geoffroy used to group arthropods and mollusks into the same branch, something that Cuvier could not accept.

8.7. Conclusion

After Lamarck and Cuvier died, the life sciences in France were influenced by the legacy of their respective schools of thought for a few decades. With respect to transformism, *Zoological Philosophy* met with notable posthumous success, and the theories of Étienne Geoffroy Saint-Hilaire were promoted and revisited in the form of "limited" transformism by his own son, Isidore.

It was at this time that Charles Darwin, a young English naturalist, set sail in 1835 on a voyage around the world, during which he collected a wealth of facts that would decisively fuel his subsequent reflection. Although he found little interest in Lamarck's ideas, which he read during his youth, he was nevertheless influenced by 18th-century thinking, particularly that of his own grandfather, Erasmus Darwin, a physician and poet who published his thoughts on living things in a volume entitled *Zoonomia* (Darwin 1794–1797). Erasmus repeated his ideas in the verses of his poem *The Temple of Nature* (1803) and even inscribed the motto *E conchis omnia* ("Everything comes from shells") on his carriage; these were all ways for him to spread his materialist and evolutionist ideas, which earned him many enemies. These few lines from *Zoonomia* illustrate the transformist thinking that was occasionally expressed at the end of the 18th century:

> From thus meditating on the great similarity of the structure of the warm-blooded animals, and at the same time of the great changes they

undergo both before and after their nativity; and by considering in how minute a portion of time many of the changes of animals above described have been produced; would it be too bold to imagine, that in the great length of time, since the earth began to exist, perhaps millions of ages before the commencement of the history of mankind, would it be too bold to imagine, that all warm-blooded animals have arisen from one living filament, which The Great First Cause endued with animality, with the power of acquiring new parts, attended with new propensities, directed by irritations, sensations, volitions, and associations; and thus possessing the faculty of continuing to improve by its own inherent activity, and of delivering down those improvements by generation to its posterity, world without end!

8.8. References

Becquemont, D. (2009). *Charles Darwin 1837-1839 : aux sources d'une découverte*. Editions Kimé, Paris.

Bonnet, C. (1745). *Traité d'insectologie ou observations sur les pucerons*. Durand, Paris.

de Buffon, G.-L.L. (1988). *Les époques de la nature*, volume 10. Mémoires du Muséum National d'Histoire Naturelle/Sciences de la Terre, Paris.

de Buffon, G.-L.L. (2007). *Buffon. Œuvres. Textes choisis*. Gallimard/Bibliothèque de La Pléiade, Paris.

Cuvier, G. (1812). *Recherches sur les ossements fossiles de quadrupèdes où l'on rétablit les caractères de plusieurs espèces d'animaux que les révolutions du globe paroissent avoir détruites*, volume I. Deterville, Paris.

Cuvier, G. (1830). *Discours sur les révolutions de la surface du globe et sur les changements qu'elles ont produits dans le règne animal*, 6th edition. Edmond d'Ocagne, Paris.

Darwin, E. (1803). *The Temple of Nature*. J. Johnson, London.

Diderot, D. (1994). *Éléments de physiologie, Philosophie*, volume I. Robert Laffont, Paris.

Gayon, J. (2006). Hérédité des caractères acquis. In *Lamarck : philosophe de la nature*, Corsi, P., Gayon, J., Gohau, G., Tirard, S. (eds). Presses Universitaires de France, Paris.

Gohau, G. (1990). *Une histoire de la géologie*. Le Seuil, Paris.

Lamarck, J.-B. (1797). *Mémoires de physique et d'histoire naturelle, établis sur les bases de raisonnement indépendantes de toute théorie ; avec l'explication de nouvelles considérations sur la cause générale des dissolutions ; sur la matière de feu ; sur la couleur des corps ; sur la formation des composés ; sur l'origine des minéraux, et sur l'organisation des corps vivans, lus à la première classe de l'Institut national dans ses séances ordinaires*. Chez l'Auteur, Paris.

Lamarck, J.-B. (1801). *Système des Animaux sans vertèbres, ou Tableau général des classes, des ordres et des genres de ces animaux ; présentant leurs caractères essentiels et leur distribution d'après les considérations de leurs rapports naturels et de leur organisation, et suivant l'arrangement établi dans les galeries du Muséum d'Histoi-re naturelle, parmi leurs dépouilles conservées ; précédé du Discours d'ouverture de l'an VIII de la République*. Deterville, Paris.

Lamarck, J.-B. (1802a). *Recherches sur l'organization des corps vivants et particulière-ment sur son origine, sur la cause de ses développemens et des progrès de sa com-position, et sur celle qui, tendant continuellement à la détruire dans chaque individu, amène nécessairement sa mort ; Précédé du Discours d'ouverture du cours de Zoologie, donné dans le Muséum National d'Histoire Naturelle, l'an X de la République*. Librairie Maillard, Paris.

Lamarck, J.-B. (1802b). *Hydrogéologie ou recherches sur l'influence qu'ont les eaux sur la surface du globe Terarestre*. Muséum d'histoire naturelle, Paris.

Lamarck, J.-B. (1809). *Philosophie zoologique, ou Exposition des considérations relatives à l'histoire naturelle des animaux*, volume 1. Chez Dentu, Paris.

Lamarck, J.-B. (1815–1822). *Histoire naturelle des animaux sans vertèbres…*, 7 volumes. Deterville, Paris.

Le Guyader, H. (1998). *Geoffroy Saint-Hilaire. Un naturaliste visionnaire*. Belin, Paris.

de Maillet, B. (1984). *Telliamed ou entretiens d'un philosophe indien avec un missionnaire français*. Fayard, Paris.

de Maupertuis, P.-L.M. (1997) *Vénus physique*. Diderot, Paris.

Roger, J. (1989). *Buffon*. Fayard, Paris.

Schmitt, S. (2023). L'édition critique des œuvres complètes de Buffon : entretien avec Stéphane Schmitt. *Revue d'histoire des sciences*, 76(1), 173–187.

Tirard, S. (2006). Génération spontanée. In *Lamarck : philosophe de la nature*, Corsi, P., Gayon, J., Gohau, G., Tirard, S. (eds). Presses Universitaires de France, Paris.

Voltaire (1964). *Chaînes des êtres. Dictionnaire philosophique, ou la raison par l'alphabet*. Garnier/Flammarion, Paris [Original published 1769].

List of Authors

Céline CHERICI
CHSSC
Université de Picardie Jules Verne
Amiens
France

Jean-Claude DUPONT
CHSSC
Université de Picardie Jules Verne
Amiens
France

Jean-Luc GUICHET
CERCLL
Université de Picardie Jules Verne
Amiens
France

Olivier PERRU
S2HEP
Université Claude Bernard Lyon 1
Lyon
France

Paolo QUINTILI
Università degli Studi di Roma Tor Vergata
Rome
Italy

Jean François THURLOY
CHSSC
Université de Picardie Jules Verne
Amiens
France

Stéphane TIRARD
CFV
Nantes Université
Nantes
France

Index of Names

C

D

E

F

G

H

I, J

K, L

M

N

P

R

S

T

U, V

W

Z

Index of Terms

A, B

C, D

E, F

G, H

I, L

M, N

O, P

S, T

V, Z

Summary of Volume 1

Chapter 3. Medicine and Healing in Early China

Dolly YANG

Chapter 4. India's Early Medical History: Buddhism and the Beginnings of Ayurveda

Kenneth G. ZYSK

Chapter 5. Greek Medicine: A Brief Historical Outline
Lorenzo PERILLI

Chapter 6. Medicine in Ancient Greece and Rome: Social Structures and Institutions
Vivian NUTTON

Summary of Volume 3

Chapter 5. The History of the Metabolism Concept: From Its Beginnings to Maturity

Jean-Claude DUPONT

Chapter 6. The History of the Gene

Antonine NICOGLOU

Chapter 9. Human Brains: Representations, Explorations and Stimulations

Mathilde LANCELOT

Chapter 10. Modern Medicine and the Sciences: A History of Ruptures

Mathieu ARMINJON

Printed and bound by CPI Group (UK) Ltd, Croydon, CR0 4YY

13/09/2026

14961072-0005